SPACE AND TIME IN CONTEMPORARY PHYSICS

An Introduction to the Theory of Relativity and Gravitation

SPACE AND TIME IN CONTEMPORARY PHYSICS
An Introduction to the Theory of Relativity and Gravitation

MORITZ SCHLICK

Rendered into English by
HENRY L. BROSE

With an Introduction by
F. A. LINDEMANN

DOVER PUBLICATIONS, INC.
Mineola, New York

Bibliographical Note

This Dover edition, first published in 1963 and republished in 2005, is an unabridged and unaltered republication of the third edition, published by Oxford University Press, New York, in 1920.

Library of Congress Cataloging-in-Publication Data

Schlick, Moritz, 1882–1936.
  [Raum und Zeit in der gegenwärtigen Physik. English]
  Space and time in contemporary physics : an introduction to the theory of relativity and gravitation / Moritz Schlick ; rendered into English by Henry L. Brose ; with an introduction by F. A. Lindemann.
    p. cm.
  "This Dover edition, first published in 1963 and republished in 2005, is an unabridged and unaltered republication of the third edition, published by Oxford University Press, New York, in 1920"—T.p. verso.
  Includes bibliographical references and index.
  ISBN 0-486-44283-7 (pbk.)
  1. Space and time. 2. Relativity (Physics) I. Title.

QC173 .59 .S65S3513 2005
530.11—dc22

2004059121

Manufactured in the United States of America
Dover Publications, Inc., 31 East 2nd Street, Mineola, N.Y. 11501

INTRODUCTION

PROBABLY no physical theory in recent times has given rise to more discussion amongst philosophers than the principle of relativity. One school of thought agrees that physicists may well be led to recast their notions of space and time in the light of experimental results. Another school, however, is of opinion that these questions are no concern of the physicists, who should make their theories fit the philosophers' conceptions of these fundamental units.

The theory of relativity consists of two parts, the old special theory, and the more recent general theory.

The main philosophic achievement of the *special* theory of relativity is probably the recognition that the description of an event, which is admittedly only perfect if both the space and time co-ordinates are specified, will vary according to the relative motion of the observer; that it is impossible to say, for instance, whether the interval separating two events is so many centimetres and so many seconds, but that this interval may be split up into length and time in different ways, which depend upon the observer who is describing it.

The reasons which force this conclusion upon the physicist may be made clear by considering what will be the impression of two observers passing one another who send out a flash of light at the moment at which they are close together. The light spreads out in a spherical shell, and it might seem obvious, since the observers are moving relatively to one another, that they cannot both remain at the centre of this shell. The celebrated Michelson-Morley experiment proves that each observer will conclude that he *does* remain at the centre of the shell. The only explanation for this is that the ideas of length and time of the one observer differ from those of the other. It is not difficult to find out exactly how much they differ, and it may be shown that there is only one set of transformations, the Lorentz-Einstein transformations, which account for the fact that each observer believes himself to be at the centre of the spherical shell. It is further a simple matter of geometry to show that these transformations are equivalent to a rotation about the axis at right angles to the relative velocity and the time. In other words, if the world is regarded as a four-dimensional space-time-manifold, the Lorentz-Einstein equations imply that each observer regards sections at right angles to his own world-line as instantaneous times. He is quite justified in doing so since the principle of relativity asserts that the space-time-manifold is homaloidal. There is no more intrinsic difference between length and time than there is between length and breadth.

The main achievement of the *general* theory of relativity has causd almost more difficulty to the school of philosophers, who would like to save absolute space and time, than the welding of space and time itself. Briefly this may be stated as the recognition of the fact that it is impossible to distinguish between a universal force and a curvature of the space-time-manifold, and that it is more logical to say the space-time-manifold is non-Euclidean than to assert that it is Euclidean, but that all our measurements will prove that it is *not*, on account of some hypothetical force. Perhaps a simple analogy may make this clearer. Suppose a golfer had always been told that all the greens were level, and had always found that a putt on a level green proceeded in a straight line. Now suppose he were playing on a strange course and found that a ball placed on the green rolled into the hole, that any putt ran in a spiral and finally reached the hole. If he were sufficiently imbued with the conviction that all greens are and must be level, he might conclude that there was some force attracting the ball to the hole. If he were of an inquiring turn of mind the golfer might try another make of ball, and possibly quite different types of balls such as tennis balls or cricket balls. If he found them all to behave in exactly the same way, though one was made of rubber, another of leather, and another filled with air, he might reasonably begin to doubt the assumption that there was a mysterious force acting on all these balls alike and begin to suspect the putting-green.

In gravitational phenomena we are confronted with an analogous case. Anywhere at a distance from matter a body set in motion continues on a straight course. In the neighbourhood of matter, however, this course is deflected. All bodies, whether large or small, dense or gaseous, behave in exactly the same way and are deflected by the same amount. Even light, which is certainly as different from matter as two things can well be, obeys the universal law. Are we not therefore bound to consider whether our space-time-manifold may not be curved rather than flat, non-Euclidean rather than Euclidean?

At first sight it might appear that there must be an easy way to settle the question. The golfer has only to fix three points on his putting-green, join them by straight lines, and measure the sum of the three angles between these lines. If the sum is two right angles the green is flat, if not, it is curved. The difficulty, of course, is to define a straight line. If we accept the definition of the shortest line, we have carried out the experiment, for the path of a ray of light is the shortest line and the experiment which determines its deflection may be read as showing that the three angles of the triangle—star—comparison star—telescope—are not equal to two right angles when the line star-telescope passes near the sun. But some philosophers appear not to accept the shortest line as the straight line. What definition they put in its place is not clear, and until they make it clear their position is evidently a weak one. It is to be hoped they will endeavour to do this, and to explain the observed phenomena rather than adopt a merely negative attitude.

This translation of Schlick's book should interest a wide circle, especially amongst those who are concerned with the general conceptions rather than the details. It would justify all, and more than all, the trouble that has been expended on it, if it served to render philosophers more conversant with the physicist's point of view and to enlist their co-operation in the serious difficulties in modern physics, which yet await solution.

F. A. LINDEMANN.

CLARENDON LABORATORY,
    OXFORD.
    March, 1920.

AUTHOR'S PREFACE TO THE SECOND EDITION

THE second edition of this book differs from the first chiefly in Chapters II and IX, which are entirely new additions. The second chapter gives a brief account of the ' special ' theory of relativity. It will probably be welcome to many readers. It seemed advisable not to assume the reader to be acquainted with the earlier theory since it has appeared that many have acquired the book, who are quite unfamiliar with the subject. The book itself gains considerably in completeness by this addition, as it now represents an introduction to the *whole* set of ideas contained in the theory of relativity, i.e. to the *special* theory as well as to the *general* theory. The beginner need not seek an entrance to the rudiments of the former from other sources.

Chapter IX of the present edition is also quite new, and cannot be omitted in a description of the fundamental notions of the theory of relativity. It develops the highly significant ideas of Einstein concerning the construction of the cosmos as a whole, by which he crowned his theory about two years ago, and which are of paramount importance for natural philosophy and for our world-view. The essential purpose of the book is to describe the physical doctrines under consideration with particular reference to their importance for our knowledge, i.e. their philosophic significance, in order that the relativity and gravitation theory of Einstein may exert the influence, to which it is justly entitled, upon contemporary thought. The fact that the second edition has rapidly succeeded the first is welcomed as an indication of a general wish to imbibe the new ideas and to strive to digest them. The book again offers its help in this endeavour. May it be of service in bringing this goal ever nearer.

I owe Professor Einstein my hearty thanks for giving me many useful hints as in the first edition.

MORITZ SCHLICK.

ROSTOCK, *January* 1919.

PREFACE TO THE THIRD EDITION

SINCE the appearance of the second edition the physical theory which is expounded in the book has been brilliantly confirmed by astronomical observations (*v.* page 65). General interest has been excited to a high degree, and the name of its creator shines with still greater lustre than before. The fundamental importance of the theory of relativity is beginning to be recognized more and more on all sides, and there is no doubt but that, before long, it will become an accepted constituent of the scientific world-view. The number of those who are filled with wonder at this achievement of genius has increased much more rapidly than the number of those who thoroughly understand it. For this reason, the demand for explanations of the underlying principles of the theory has not decreased but, on the contrary, is growing. This is shown by the fact that the second edition, although more numerous than the first, became exhausted more rapidly.

The present edition varies from the previous one only in small additions and other slight improvements. I have endeavoured to meet the wishes which observant readers have expressed to me personally or in writing. I hope that the book will now somewhat better fulfil its good purpose of leading as far as possible into the wonderful thought-world of the theory of relativity. Among those to whom I am indebted for suggestions, I wish to express my special thanks to Professor E. Cohn, of Strassburg (now at Rostock).

MORITZ SCHLICK.

ROSTOCK, *January* 1920.

BIBLIOGRAPHICAL NOTE

REFERENCE may be made to the following books dealing with the general theory of relativity:

A. S. Eddington. *Report on the Relativity Theory of Gravitation.* Fleetway Press.

A. S. Eddington. *Space, Time, and Gravitation.* Camb. Univ. Press. (In the Press.)

An elementary account is given in:

Erwin Freundlich. *The Foundations of Einstein's Theory of Gravitation* (trans. by Henry L. Brose). Camb. Univ. Press.

Albert Einstein. *The Special and General Theory of Relativity* (trans. by R. W. Lawson). Messrs. Methuen.

Henry L. Brose. *The Theory of Relativity.* An Essay. B. H. Blackwell, Oxford.

The most important German book on the subject is:

Hermann Weyl. *Raum, Zeit und Materie.* Jul Springer, Berlin. This gives all the details of the mathematical reasoning.[1]

Einstein's epoch-making papers are:

'Grundlagen der allgemeinen Relativitätstheorie.' *Ann. d. Physik,* 4. Folge. Bd. 49, S. 769.[2]

Grundlagen des allgemeinen Relativitätsprincips. J. A. Barth. Leipzig. 1916.

'Erklärung der Perihelbewegung des Merkur aus der allgemeinen Relativitätstheorie.' *Sitzungsberichte der königl. preuss. Akad. der Wissenschaften,* Nov. 1915. Bd. xlvii.

[1]Published by Dover in English as *Space, Time, Matter.*
[2]Included in Dover's *The Principle of Relativity* by Einstein and others.

The evolution of the ideas which are discussed in Chapter X of the present book may be traced in the following works, in addition to those mentioned in the text:

Jevons. The *Principles of Science*. Macmillan & Co.

H. Poincaré. *La Valeur de la Science*. Paris.—*La Science et l' Hypothèse*.[3] Paris.

Ernest Mach. *Erkenntnis und Irrtum*. Leipzig.—*Die Analyse der Empfindungen*.[4]

Jos. Petzoldt. *Das Weltproblem*. Leipzig.

Aloys Müller. *Das Problem des absoluten Raumes und seine Beziehung zum allegemeinen Raumproblem*. Vieweg, Braunschweig.

Moritz Schlick. *Allegemeine Erkenntnislehre*. Jul. Springer, Berlin.

I wish to take this opportunity of thanking Mr. J. W. N. Smith, M.A., of Christ Church (now at Rugby) for the great care he has taken in revising the proof-sheets. Professor Schlick and Dr. Wichmann kindly compared the translation with the original, and made a number of helpful suggestions. I am indebted to Miss Olwen Joergens for the English rendering of the quotation from Giordano Bruno. Professor Mitchell, Vice-Chancellor of Adelaide University, kindly verified the philosophical terminology of the last chapter.

HENRY L. BROSE.

CHRIST CHURCH, OXFORD,
March, 1920.

[3] Published by Dover in English as *The Value of Science*. and *Science and Hypothesis*.

[4] Published by Dover in English as *The Analysis of Sensations*.

CONTENTS

SPACE AND TIME IN CONTEMPORARY PHYSICS

An Introduction to the Theory of
Relativity and Gravitation

I

FROM NEWTON TO EINSTEIN

At the present day physical research has reached such a degree of generalization of its first principles, and its standpoint has attained to such truly philosophic heights, that all previous achievements of scientific thought are left far behind. Physics has ascended to summits hitherto visible only to *philosophers,* whose gaze has, however, not always been free from metaphysical haziness. Albert Einstein is the guide who has directed us along a practicable path leading to these summits. Employing an astoundingly ingenious analysis, he has purged the most fundamental conceptions of natural science by removing all the prejudices which have for centuries past remained undetected in them: thus revealing entirely new points of view, and building up a physical theory upon a basis which can be verified by actual observation. The fact that the refinement of the conceptions, by a critical examination of them from the viewpoint of the theory of knowledge, is simultaneously combined with the physical application which immediately made his ideas experimentally verifiable, is perhaps the most noteworthy feature of his achievement: and it would be remarkable, even if the problem with which he was able to grapple by using these weapons had not happened to be gravitation—that riddle of physics which so obstinately resisted all efforts to read it, and the solution of which must of necessity afford us glimpses into the inner structure of the universe.

The most fundamental conceptions in physics are those of Space and Time. The unrivalled achievements in research, which in past centuries have enriched our knowledge of physical nature, left these underlying conceptions untouched until the year 1905. The efforts of physicists had always been directed solely at the substratum which *occupied* space and time: they had taught us to know, more and more accurately, the constitution of matter and the law of events which occurred *in vacuo,* or as it had, till recently, been expressed, in the 'aether'. Space and Time were regarded, so to speak, as vessels containing this substratum and furnishing fixed systems of reference, with the help of which the mutual relations between bodies and events had to be determined: in short, they actually played the part which Newton had set down for them in the well-known words: 'Absolute, true and mathematical time flows in virtue of its own nature uniformly and without reference to any external object'; and 'absolute space, by virtue of its own nature and without reference to any external object, always remains the same and is immovable'.

From the standpoint of the theory of knowledge, the objection was quite early raised against Newton, that there was no meaning in the terms Space and Time as used without 'reference to an object'; but, for the time being, physics had no cause to trouble about these questions: it merely sought to explain observed phenomena in the usual way, by refining and modifying its ideas of the constitution and consistent behaviour of matter and the 'aether'.

An example of this method is the hypothesis which was put forward by H. A. Lorentz and Fitzgerald, that every body which is in motion relatively to the aether is subject to a definite contraction along the direction of motion (the socalled Lorentz-contraction), which depends upon the velocity of the body. This hypothesis was set up in order to explain why it seemed impossible to detect 'absolute' rectilinear motion of our instruments by means of the experiment of Michelson and Morley (which will be discussed below), whereas, according to the prevalent physical ideas of the time, this should have been possible. The whole trend of physical discovery made it evident that this hypothesis would not be permanently satisfactory (as we shall see immediately), and this meant that the time was come when the consideration of motion in physics had to be founded on reflections of a philosophic nature. For Einstein recognized that there is a much simpler way of explaining from first principles the negative result of Michelson and Morley's experiment. No special physical hypothesis at all is required. It is only necessary to recognize the principle of relativity, according to which a rectilinear uniform 'absolute' motion can never be detected, and the fact that the conception of motion has only a physical meaning when referred to a material body of reference. He saw also that a critical examination of the assumptions upon which our space- and time-measurements have hitherto been tacitly founded is necessary. Amongst these unnecessary and unwarrantable assumptions were found, e.g. those which concerned the absolute significance of such space- and time-conceptions as 'length', 'simultaneity', &c. If these assumptions are dropped, the result of Michelson and Morley's experiment appears self-evident, and on the ground thus cleared is constructed a physical theory of wonderful completeness, which develops the consequences of the above fundamental principle; it is called the 'special theory of relativity, because, according to it, the relativity of motions is valid only for the special case of uniform rectilinear motion.

The special principle of relativity indeed takes one considerably beyond the Newtonian conceptions of Space and

Time (as will be seen from the short account in the next chapter), but does not fully satisfy the philosophic mind, inasmuch as this restricted theory is only valid for uniform rectilinear motions. From the philosophic standpoint it is desirable to be able to affirm that *every* motion is relative, i.e. not the particular class of uniform translations only. According to the special theory, irregular motions would still be absolute in character; in discussing them we could not avoid speaking of Space and Time 'without reference to an object'.

But since the year 1905, when Einstein set up the special principle of relativity for the *whole* realm of physics, and not for mechanics alone, he has striven to formulate a generalized principle which is valid not only for uniform rectilinear motions, but also for any arbitrary motion whatsoever. These endeavours were brought to a happy conclusion in 1915, being crowned with complete success. They led to such an extreme degree of relativization of all space- and time-determinations that it seems impossible to extend it any further; these space- and time-determinations will henceforth be inseparably connected with matter, and will have meaning only when referred to it. Moreover, they lead to a new theory of gravitational phenomena which takes physics very far beyond that of Newton. Space, time, and gravitation play in Einstein's physics a part fundamentally different from that assigned to them by Newton.

The importance of these results, in their bearing upon the underlying principles of natural philosophy, is so stupendous that even those who have only a modest interest in physics or the theory of knowledge cannot afford to pass them by. One has to delve deep into the history of science to discover theoretical achievements worthy to rank with them. The discovery of Copernicus might suggest itself to the mind; and if Einstein's results do not exert as great an influence on the world-view of people in general as the Copernican revolution, their importance as affecting the purely theoretical picture of the world is correspondingly greater, inasmuch as the deepest foundations of our knowledge concerning physical nature have to be remodelled much more radically than after the discovery of Copernicus.

It is therefore easy to understand, and gratifying to note, that there is a general desire to penetrate into this new field of thought. Many are, it is true, repelled by the external form of the theory, because they cannot acquire the highly complicated mathematical technique which is necessary for an understanding of Einstein's researches: but the wish to be initiated into these new views, even without this technical help, must be satisfied, if the theory is to exercise its rightful influence in forming the modern view of the world. And it can be satisfied without difficulty, for the principles are as simple as they are profound. The conceptions of Space and Time were not in the first place evolved by a complicated process of scientific thinking, but we are compelled to use them incessantly in our daily life. Starting from the most familiar conceptions of everyday life, we can proceed step by step to exclude all arbitrary and unjustified assumptions, until we are finally left with Space and Time in the simple form in which they play their part in Einstein's physics. We shall adopt this plan here, in order to crystallize the fundamental ideas in particular of the new theory of Space. We get them without any effort, by merely expelling from the traditional notion of Space all ambiguities and unnecessary thought-elements. We shall clear a way leading to the general theory of relativity, if we get our ideas of Space and Time precise by subjecting them to a critical examination, inasmuch as they serve as a foundation for the new doctrine and make it intelligible. We shall prepare ourselves for this task by considering first the thoughts underlying the 'special' theory of relativity.

II

THE SPECIAL PRINCIPLE OF RELATIVITY

Michelson and Morley's experiment forms the best introduction to this principle, both historically and for its own sake. Historically, because it gave the first impulse towards setting up the relativity-theory; and in itself, because the suggested explanations of the experiment bring the old and new currents of thought into strongest relief with one another.

The condition of affairs was as follows. The electromagnetic waves, of which light is composed, and which propagate themselves with a velocity c equal to 300,000 kilometres per second (186,000 miles per sec.), were regarded by the older physicists as changes of state, transmitted as a wave-disturbance in a substance called 'aether', which completely filled all empty space, including even that between the smallest particles of material bodies. Accordingly, light would be transmitted relatively to the aether with the above velocity c (i.e. one would obtain the value 300,000 kilometres per second) if the velocity were measured in a co-ordinate system, fixed in the aether. If, however, the velocity of light were to be measured from a body which was moving relatively to the aether with the velocity q in the direction of the light-rays, the observed velocity of the light-rays should be $c - q$, for the light waves would hurry past the observer more slowly since he is moving with them in their direction. If he were moving directly towards the waves of light, he should get $c + q$ for its velocity by measurement.

But, so the argument continues, we on the earth are exactly in the position of the observer moving relatively to the aether: for numerous observations had compelled us to assume that the aether does not partake of the motion of bodies moving through it, but preserves its state of undisturbed rest. This means that our planet, our measuring instruments, and all other things on it, rush through the aether, without in the slightest dragging it along with them; it slips through all bodies with infinitely greater ease than the air between the planes of a flying machine. Since the aether is nowhere in the world to take part in any motion of such bodies, a co-ordinate-system which is stationary in it fulfils the function of a system which is 'absolutely at rest'; and there would thus be meaning in the phrase 'absolute motion' in physics. This would indeed not be absolute motion in the strictly philosophical sense, for we should understand it as a motion relative to the aether, and we could still ascribe to the aether and the cosmos embedded in it any arbitrary motion or rest in 'space'—but the possibility is quite devoid of meaning, as we should no longer be dealing with observable quantities. If there is an aether, the system of reference which is fixed, i.e. at rest, in it must be unique amongst all others. The proof of the physical reality of the aether would necessarily, and could only, consist in discovering this unique system of reference. For example, we might show that only with reference to this system is the velocity of transmission of light the same in all directions, viz. *c*, and that this velocity is different when measured relatively to other bodies.—After what has been said, it is clear that this unique system, which is absolutely at rest, could not be moving with the earth, for the earth traverses about 30 kilometres per second in its course round the sun. Our instruments thus move with this velocity relative to the aether (if we neglect the velocity of the solar system, which would have to be added to this). This velocity of 30 kilometres per second—for a first approximation we may suppose it to be uniform and rectilinear—is indeed small in comparison with c; but, with the help of a sufficiently delicately arranged experiment, it should be possible to measure a change of this order in the velocity of light, without difficulty. Such an experiment was devised by Michelson and Morley. It was carefully arranged in such a way that even the hundredth part of the expected amount could not have escaped detection if it had been present.

But no trace of a change was to be found. The principle of the experiment consisted in a ray of light being reflected to and fro between two fixed mirrors placed opposite to one another, the line joining the centres of the mirrors being in one case parallel to the earth's motion, and in another perpendicular to it. An easy calculation shows that the time taken by the light to traverse the space between the two mirrors (once to and fro) is in the second case only $\sqrt{1 - q^2/c^2}$ of the value obtained in the first case, if q denotes the velocity of the earth relatively to the aether. The absence of any change, in the initial interference fringes, proves with great accuracy that the time taken is exactly the same in both cases.

Hence the experiment teaches us that light also propagates itself in reference to the earth with equal velocity in all directions, and that we cannot detect ' absolute ' motion (i.e. motion with respect to the aether) by this means.

The same result holds for other methods; for, besides Michelson and Morley's attempt, other experiments (for instance, that of Trouton and Noble concerning the behaviour of a charged condenser) have led to the conclusion that absolute motion (we are throughout these remarks only speaking of uniform rectilinear motion) cannot be established in any way.

This fact seemed new as far as optical and other electromagnetic experiments were concerned. It had long been known, on the other hand, that it was impossible to detect any absolute rectilinear uniform motion by means of *mechanical* experiments. This principle had been clearly stated in Newtonian mechanics. It is a matter of everyday experience that all mechanical events take place in a system which is moving uniformly and rectilinearly (e.g. in a moving ship or train) exactly in the same way as in a system which is at rest relatively to the earth. But for the inevitable occurrence of jerks and rocking (which are *non*-uniform motions) an observer enclosed in a moving air-ship or train could in no wise establish that his vehicle was moving.

To this old theorem of mechanics there was now to be added the corollary that electrodynamical experiments (which include optical ones) give an observer no indication as to whether he and his apparatus are at rest or moving uniformly and rectilinearly.

In other words, experience teaches us that the following theorem holds for all physics: 'All laws of physical nature which have been formulated with reference to a definite coordinate system are valid, in precisely the same form, when referred to another co-ordinate system which is in uniform rectilinear motion with respect to the first.' This empirical law is called the 'special theory of relativity', because it affirms the relativity of uniform translations only, i.e. of a very special class of motions. All physical events take place in any system in just the same way, whether the system is at rest or whether it is moving uniformly and rectilinearly. There is no absolute difference between these two states; I may regard the second equally well as being that of rest.

The empirical fact of the validity of the special principle of relativity, however, entirely contradicts the considerations made above concerning the phenomenon of light-transmission, as founded upon the aether-theory. For, according to the latter, there should be one unique system of reference (that which is fixed with reference to the ' aether '), and the value obtained for the velocity of light should have been dependent upon the motion of the system of reference used by the observer. Physicists were confronted with the difficult problem of explaining and disposing of this fundamental contradiction; this is the point of divergence of the old and the new physics.

H. A. Lorentz and Fitzgerald removed the difficulty by making a new physical hypothesis. They assumed that all bodies, which are put in motion with reference to the aether, suffer a contraction to $\sqrt{1 - q^2/c^2}$ of their length in the direction of their motion. Hereby the negative result of Michelson and Morley's experiment would in fact be completely explained; for, if the line between the two mirrors used for the purpose were to shorten of its own accord as soon as it is turned so as to be in the direction of the earth's motion, light would take less time to traverse it, and indeed, the reductions would be exactly the amount given above (viz. that by which the time of passage should have been greater than in the position perpendicular to the earth's motion). The effect of the absolute motion would thus be exactly counterbalanced by this Lorentz-Fitzgerald contraction; and, by means of similar hypotheses, it would also be possible to give a satisfactory account of Trouton and Noble's condenser experiment and other experimental facts.

We thus see that, according to the point of view just described, there is actually to be an absolute motion in the physical sense of the term (viz. with reference to a material aether); but, since such a motion cannot be observed in any way, special hypotheses are devised to explain why it always eludes our perception. In other words, according to this view the principle of relativity does *not* hold, and the physicist is obliged to explain, by means of special hypotheses, why all physical phenomena in spite of this take place actually *as if* it did hold. An aether is really to exist, although a unique body of reference of this kind nowhere manifests itself.

In opposition to this view, modern physics, following Einstein, asserts that, since experience teaches us that the special principle of relativity actually holds, it is to be regarded as a *real* physical law; since, furthermore, the aether as a substance obstinately evades all our attempts at observing it, and all phenomena occur as if it did not exist, the word 'aether' lacks physical meaning, and therefore aether does not exist. If the principle of relativity and the non-existence of the aether cannot be brought to harmonize with our previous arguments about the transmission of light, these arguments must clearly be reconsidered and revised. It is to Einstein that the credit falls of discovering that such a revision is possible, viz. that these arguments are based on assumptions concerning the measurement of space and time which have not been tested, and which we only require to discard in order to do away with the contradiction between the principle of relativity and our notions about the transmission of light.

Thus, if an event propagates itself, with respect to a coordinate system K, in any direction with the velocity c, and if a second system K^1 move relative to K in the same direction with the velocity q, the velocity of transmission of the event as viewed from the system K^1 is of course only equal to $c-q$, *if* it is assumed that distances and times are measured in the two systems with the same measuring units. This assumption had hitherto been tacitly used as a basis. Einstein showed that it is in no wise self-evident: that one could with equal right (indeed with greater right, as the results will show) put the value for the velocity of transmission in both systems equal to c; and that the lengths of distances and of times then have different values for different systems of reference moving with reference to one another. The length of a rod, the duration of an event, are not absolute quantities, as was always assumed in physics before the advent of Einstein, but are dependent on the state of motion of the co-ordinate system in which they are measured. The methods which are at our disposal for measuring distances and times yield different values in systems which are in motion relatively to one another. We shall now proceed to explain this more clearly.

For the purpose of 'measurement', i.e. for the quantitative comparison of lengths and times, we require measuring-rods and clocks. Rigid bodies, the size of which we assume to be independent of their position, serve as measuring-rods; the term clock need not necessarily be confined to the familiar mechanical object, but may denote any physical contrivance which exactly repeats the same event periodically; e.g. light-vibrations may serve as a clock (this was the case in Michelson and Morley's experiment).

No essential difficulty arises in determining a moment or the duration of an event, if a clock is at our disposal at the place where the event is happening; for we need only note the reading of the clock at the moment the event under observation begins, and again at the moment it ceases. The sole assumption we make is that the conception of the 'simultaneity (time-coincidence) of two events occurring at the same place' (viz. the reading of the clock and the beginning of the event) has an absolutely definite meaning. We may make the assumption, although we cannot define the conception or express its content more clearly; it belongs to those ultimate data, which become directly known to us as an experience of our consciousness.

The position is different, however, when we are dealing with two events which occur at *different* places. To compare these events in point of time, we must erect a clock at each place, and bring these two clocks into agreement with one another, viz. regulate them so that they beat synchronously, i.e. give the same reading at the 'same moment'. This regulation, which is equivalent to establishing the conception of simultaneity for different places, requires a special process. We are obliged to resort to the following method. We send a light-signal from the one clock placed at *A* (let us say) to the second at *B*, and reflect it thence back to *A*. Suppose that, from the moment of sending to that of receiving the signal, the clock *A* has run on for two seconds, then this is the time which the light has required to traverse the distance *AB* twice. Now since (according to our postulate) light propagates itself in all directions with the same velocity *c*, it takes just as long for the initial as for the return journey, i.e. one second for each. If we now emit a light-signal in *A* at precisely twelve o'clock, after having arranged with an observer in *B* to set his clock at one second past 12 o'clock when he receives the signal, then we shall rightly consider that we have solved the problem of synchronizing the two clocks. If there are other clocks at other places, and if we bring them all into agreement with the one at *A* according to the method described for *B*, then they will agree amongst themselves if compared by the same process. Experience teaches us that the only time-data which do not lead to contradictions are those which are got by using signals which are independent of matter, i.e. are transmitted with the same velocity through a vacuum. Electro-magnetic waves travelling with the speed of light fulfil this condition.

If we were to use sound-signals in the air, for instance, the direction of the wind would have to be taken into account. The velocity of light c thus plays a unique part in Nature.

Hitherto we have assumed that the clocks are at rest relatively to one another and to a fixed body of reference K (as the earth). We shall now suppose a system of reference K^1 (e.g. a railway train travelling at an enormous rate) moving relatively to K with the velocity q in the direction of A to B. The clocks at different points in K^1 are to be supposed regulated with one another in exactly the same way as was just described for those in K. K^1 may for this purpose be considered to be at rest equally well as K, when its clocks were regulated. What happens when observers in K and K^1 attempt to get into communication with one another?

Suppose a clock A^1 at rest in K^1 to be in immediate proximity to the clock A at rest in K, at precisely the moment at which both clocks A and A^1 indicate 12; and suppose a second clock B^1 at rest in K^1 to be at the place B, whilst the corresponding clock at rest in K at the same place indicates 12. An observer on K will then say that A^1 coincides with A at the same moment, i.e. simultaneously (at exactly 12 o'clock) when B^1 coincides with B. At the moment when the coincident clocks A and A^1 both indicate 12, let a light-signal flash out from their common position. The rays reach B when the clock at B indicates one second past 12; but the clock B^1, being on the moving body K^1, has moved away from B a distance q, and will have moved slightly further away before it is reached by the light-signal. This means that, for an observer at rest on K, the light takes longer than one second to travel from A^1 to B^1. It will now be reflected at B^1, and will arrive back at A^1 in *less* than one second, since A^1, according to the observer in K, moves towards the light. This observer will therefore conclude that the light takes longer to traverse the distance from A^1 to B^1 than that from B^1 to A^1: since in the first case B^1 hastens away from the light-ray, whereas in the second case A^1 goes to meet it. An observer in K^1, however, judges otherwise. Since he is at rest relatively to A^1 and B^1, the times taken by the signal to travel from A^1 to B^1, and thence back from B^1 to A^1, are exactly the same: for, with reference to his system K^1, light propagates itself with equal velocity c in both directions (according to the postulate we have established on the basis of Michelson and Morley's result).

We thus arrive at the conclusion that two events, which are of *equal* duration in the system K^1, occupy *different* lengths of time when measured from the system K. Both systems accordingly use a different time-measure; the conception of duration has become relative, being dependent on the system of reference, in which it is measured. The same holds true, as immediately follows, of the conception of simultaneity: two events, which, viewed from one system, occur simultaneously, happen for an observer in another system at different times. In our example, when A coincides with A^1 in position, the two clocks at the common point indicate the same time as the clock B when B coincides with B^1; but the clock B^1, belonging to the system K^1, indicates a *different* time at this place. The former two coincidences are thus only simultaneous in K but not in the system K^1.

All this arises, as we see, as a necessary consequence of the regulation of clocks, which was founded upon the principle that light always transmits itself with constant velocity: no other means of regulation is possible without introducing arbitrary assumptions.

We also obtain different values for the lengths of bodies

taken along the direction of motion, if they are measured from different systems. This is immediately evident from the following. If I happen to be at rest in a system K, and wish to measure the length of a rod AB which is moving with reference to K in the direction of its own axis, I must either note the time that the rod takes to move past a fixed point in K, and multiply this time by the velocity of the rod relative to K (by doing which we should find the length to be dependent on the velocity, on account of the relativity of duration); or I could proceed to mark on K at a definite moment two points P and Q, which are occupied by the two ends A and B respectively at that precise moment, and then measure the length of PQ in K. Since simultaneity is a relative conception, the coincidence of A with P, if I make observations from a system moving with the rod, will not be simultaneous with the coincidence of B with Q: but at the time that A coincides with P, the point B will, for me, be at a point Q^1 slightly removed from Q, and I shall regard the distance PQ^1 as the true length of the rod. Calculation shows that the length of a rod, which has a value a in a system with reference to which it is at rest, assumes the value $a \sqrt{1 - q^2/c^2}$ for a system which is moving relatively to it with the velocity q. This is precisely the Lorentz-contraction. It no longer appears as a physical effect brought about by the influence of 'absolute motion', as was the case according to Lorentz and Fitzgerald, but is merely the result of our methods of measuring length and times. The question which is often put forward by the beginner, as to what the 'real' length of a rod is, and whether it 'really' contracts on being moved, or whether the change in length is only an apparent one—is suggested by a misunderstanding. The diverse lengths, which are measured in various systems moving with uniform motion relatively to one another, all 'really' belong to the rod equally; for all such systems are equivalent.   No contradiction is contained in this, since 'length' is only a *relative* conception.

The conceptions 'more slowly' and 'more quickly' (not only 'slowly' and 'quickly') are, according to the new theory, relative.   For, if an observer in K always compares his clock with the one in K^1, which he just happens to be passing, he will find that these clocks lag more and more behind his own: he will hence declare the rate of the clocks in K^1 to be slower than his own.   Exactly the same, moreover, happens to the observer in K^1, if he compares his clock with the successive clocks of K which he happens to encounter.   He will assert that the clocks fixed in his own system are going at a faster rate; and this indeed with just as much right as the other had in affirming the contrary.

All these connected results can be most easily followed if they are expressed mathematically; we can then grasp them as a whole.   For this purpose we only require to set up the equations, which enable us to express the time and place of an event, referred to one system by corresponding quantities referred to the other system.   If x_1, x_2, x_3 are the space-co-ordinates of an event happening at the time t in the system K; and if $x^1{}_1$, $x^1{}_2$, $x^1{}_3$, t^1 are the corresponding quantities referred to K^1; then these equations of transformation (they are termed the 'Lorentz-transformation') enable us to calculate the quantities $x^1{}_1$, $x^1{}_2$, $x^1{}_3$, t^1, if x_1, x_2, x_3, t are given and vice versâ.   (For further details see the references at the end of this book.)

Such are, in a few words, the main features of the kinematics of the special theory of relativity.   Its great importance in physics is derived from the electro-dynamics and mechanics which correspond to this type of kinematics. But for our present purpose it is not necessary to go into greater detail.   We shall only mention one extraordinary result.

Whereas in the older physics the law of Conservation of Energy and that of Conservation of Mass existed entirely unrelated, it has been shown that the second law is no longer strictly in agreement with the former, and must therefore be abandoned. Theory leads to the following view. If a body take up an amount of energy E (measured in a system which is at rest with reference to that of the body), the body behaves *as if* its mass were increased by the amount $\dfrac{E}{c^2}$. That is, we cannot say that each body has a constant factor m which has the significance of a mass independent of its velocity. If, now, the quantity $\dfrac{E}{c^2}$ is to be regarded as an *actual* increase of mass, i.e. if energy has the property of inertia, it is an obvious step not only to trace the increase of mass back to an increase of energy but also to regard the inertial mass m as being dependent upon a quantity of energy $E = mc^2$ contained by the body. This amount is very great owing to the enormous value of c, the velocity of light. This assumption is in very good agreement with the enormous store of internal energy of the atom, as deduced from recent researches. Physics, therefore, no longer recognizes both of the above laws, but only that of the Conservation of Energy. The Principle of Conservation of Mass, which has hitherto been regarded as a distinct fundamental law of natural science, has been traced back to the Principle of Energy, and has been recognized as being only approximately true. It is found to be nearly true, inasmuch as all increases of energy which are experimentally possible are in general negligible compared with the enormous store of internal energy mc^2, so that these changes of mass are scarcely observable.

That which particularly interests us here is that the

theory of relativity entirely does away with the traditional conceptions of space and time, and banishes 'aether' as a substance out of physics. We saw earlier that the existence of such an aether implied in physical terms that a definite co-ordinate system (that which is at rest relatively to the aether) would have to be unique amongst all others, i.e. with reference to this system physical laws would assume a particular form. As our theory allows no such unique system, and since, on the contrary, all systems which have a uniform translation with regard to one another are equivalent, the belief in a material aether is incompatible with the principle of relativity. We may no longer regard light-waves as a change in the condition of a *substance,* in which they are propagated with the velocity c; for then this substance would have to be at rest in all equivalent systems, and that of course entails a contradiction. The electromagnetic field is, on the contrary, to be regarded as being independent and not requiring a 'carrier'. Since we are free to use words at pleasure, there is no objection to using the word 'aether' in future to represent the vacuum (empty space) with its electromagnetic field, or as endowed with the metrical properties which are to be discussed below; we must be very cautious, however, not to picture it as matter.

We thus see that, in addition to the conceptions of space and time, that of *substance* is crystallized in a purified form by the critical application of the *special* theory of relativity. This process only reaches completion, however, in the *general* theory of relativity. However great the revolution wrought by the special theory may have seemed, the claim that *all* motions without exception should be of a relative character (i.e. that only motions of bodies *relatively to one another* are to enter into physical laws) brings about such a strange world-picture and leads to such bold conclusions that, in comparison with it, the reconstruction of conceptions imposed upon us by the special theory of relativity seems modest and incomplete.

To gain an easy approach to the formidable structure of ideas contained in the general theory of relativity, we shall start afresh with quite elementary reflections and simple questionings.

III

THE GEOMETRICAL RELATIVITY OF SPACE

THE most fundamental question which may be asked concerning Space and Time is, to express it in familiar language for the present: are Space and Time actually real?

From the earliest times an inconclusive controversy was waged by the philosophers as to whether empty space, the κενόν, were real, or merely identical with nothingness. But even at the present day not every one, be he scientist, philosopher, or general reader, would straightway answer this question by a simple negative or affirmative. No one, indeed, regards Space and Time as real in quite the same sense as the chair on which I sit, or the air which I breathe. I cannot deal with space as with material objects or with energy, which I can transport from one place to another, manipulate at will, buy and sell. Every one feels that there is some difference between them; Space and Time are, in some sense or other, less *independent* than the things which exist in them; and philosophers have often emphasized this lack of independence by stating that neither exists in itself. We could not speak of Space if there were no material bodies; and the conception of Time would likewise be devoid of meaning if no events or changes took place in the world. But, even for the popular mind, Space and Time are not merely *nothing*; for are there not great departments of engineering which are wholly devoted to overcoming them?

Of course the decision of this question depends upon what is understood by 'Reality'. Now, even if this conception is difficult, perhaps even impossible, to define, yet the physicist is in the happy position of being able to satisfy himself with a definition which allows him to fix the limits of his realm with absolute certainty. 'Whatever can be measured is real.' The physicist may use this sentence of Planck's as a general criterion, and say that only that which is measurable possesses indisputable reality, or, to define it more carefully, physical objectivity.

Are Space and Time measurable? The answer seems obvious. What would indeed be measurable if it were not Space and Time? Do not our clocks and measuring-scales serve just this purpose? Is there not even a special science which is concerned with nothing else than with the measurement of space, without reference to any bodies, viz. metrical geometry?

But let us be cautious! It is known that there is difference of opinion about the nature of geometrical objects —even if this were not the case, we have recently learnt to look searchingly into the fundamental conceptions of the sciences above all for concealed or unproved premises. We shall thus have to investigate whether the current view of geometry, as a doctrine of the properties of space, is not influenced by certain unjustified notions, from which it must be released. In fact, philosophic criticism has for some time affirmed the necessity for doing so, and busied itself with the task, and has thereby already developed ideas about the relativity of all spatial relations. We may regard the space-time-view of Einstein's theory as the logical shaping and application of these ideas; a continuous path leads from them to the theory, along which the meaning of the question of the reality of Space and Time becomes ever clearer. We shall use this road as a means of access to the new ideas.

Let us begin by reflecting on a simple imaginary experiment, which almost every one who has thought about these

matters has performed mentally and which is particularly
well described by Henri Poincaré. Let us suppose that all
material bodies in the world increase enormously in size
over-night to a hundred times their original dimensions; my
room, which is to-day six metres long, would to-morrow have
a length of 600 metres. I myself should be a Goliath 180
metres high, and should be inscribing letters a metre high on
paper with a pen 15 metres long; and similarly all other
dimensions of the universe are to be supposed altered to
a like degree, so that the new world, although a hundred
times increased, would still be geometrically similar to the
old one. 'What would my impressions be in the morning,'
Poincaré asks, 'after this astonishing change?' And he
answers: 'I should not observe the slightest difference. For
since, according to our assumption, all objects, including my
own body, all measuring-scales and instruments, have shared
in this hundredfold magnification, every means of detecting
this change would be wanting; I should call the length of
my room 6 metres as before, since my metre-scale would
divide into it six times, and so on.' What is still more important, this whole alteration would exist *only* for those who
erroneously argue that Space is absolute. Truth compels us
to say that, since space is relative, no change has taken place,
and that this is the reason why we were unable to notice
anything. Thus, the universe, which we imagined magnified
a hundredfold, is not only indistinguishable from the original one; it is simply the *same* universe. There is no meaning in talking of a difference, because the absolute size of a
body is not 'real'.

The exposition of Poincaré must be carried a little further
to be quite convincing. The fiction of a universal alteration
in the size of the world, or a part of it, is devoid of any ap-
preciable meaning from the very outset, unless definite assumptions are made as to how the physical constants are to

behave in this deformation. For natural bodies have not only a geometrical shape, but they also possess physical properties, e. g. mass. If, after a hundredfold linear magnification of the world, we substitute the former values for the mass of the earth and the objects it contains, in Newton's attraction formula, we shall only get a 10,000th of the previous value for the weight of a body on the earth's surface, since this weight is inversely proportional to the square of the distance from the earth's centre. Can we not establish this change in weight, and thus arrive indirectly at the absolute increase in size? We might think that this would be possible by observations of a pendulum, for the time of vibration (period) of a pendulum would be just 1,000 times slower on account of the decrease in weight and increase in length. But would this retardation be observable? Would it possess physical reality? The question is again unanswerable, unless it is stated how the rotational velocity of the earth is affected by the deformation; for our time-measures are based upon comparison with the former.

The attempt to observe the decrease in weight by means of a spring-balance (say) would likewise be in vain; for special assumptions about the behaviour of the coefficient of elasticity of the spring would again be necessary in this supposed magnification.

The fiction of a purely geometrical deformation of all bodies is therefore entirely without significance; it has no definite physical meaning. If one fine day we were to observe a slowing down of all our pendulum-clocks, we could not infer that the universe had been magnified during the night, but the remarkable phenomenon could be explained by means of other physical hypotheses. Inversely, if I assert that all linear dimensions have been lengthened a hundredfold since yesterday, no experience could prove the contrary; I should only have to affirm at the same time that while all masses had increased a hundredfold in value, the rate of the earth's spin and of other events had, on the other hand, decreased to a hundredth of their former value. It is easily seen from the elementary formulae of Newtonian Mechanics that, with these assumptions, exactly the same numbers result from the calculations as before for all observable quantities (at least as far as gravitational and inertial effects are concerned). The change has thus no physical meaning.

From reflections of this kind, which may be multiplied at pleasure, and which are still based on Newtonian mechanics, it is already clear that space-time considerations are inseparably bound up with other physical quantities; and if we abstract some from the rest, we must by careful comparison with experience try to discover in what sense a real meaning is to be attached to the abstraction.

The reflections of Poincaré, supplemented in the manner indicated, teach us beyond doubt that we can imagine the world transformed by means of far-reaching geometrical-physical changes into a new one, which is completely indistinguishable from the first, and which is completely identical with it physically, so that the transformation would not actually signify a real happening. We started by considering the case in which the imaginary transformed world is geometrically *similar* to the original one; the conclusions drawn are not in the slightest affected by dropping this assumption. If we, for instance, assumed that the dimensions of all objects are lengthened or shortened in one direction only, say that of the earth's axis, we should again not notice this transformation, although the shape of bodies would have changed completely, spheres becoming ellipsoids of rotation, cubes becoming parallelopideds, and indeed perhaps very elongated ones. But if we wished to establish, by means of a measuring-scale, the change in length as compared with the breadth, our effort would be in vain; since our measuring-rod, when we turned it into the direction of the earth's axis for the purpose of carrying out a measurement, would, according to our supposition, be correspondingly lengthened or shortened. Nor could we become aware of the deformation directly by means of the senses of sight or touch; for our own body has likewise become deformed, as well as our eye-balls, and also the wave-surfaces of light. Again, we must conclude that there is no 'real' distinction between the two worlds; the imagined deformation is not ascertainable by any measurement, i.e. has no physical objectivity. It is easily seen that the arguments just presented may be generalized still further: we can imagine with Poincaré that the objects in the universe are arbitrarily distorted in arbitrary directions, and the distortion need not be the same for all points, but may vary from place to place.—As long as we suppose that all measuring instruments, including our own bodies with their sense-organs, share in the local deformation for each place, the whole transformation immediately becomes unascertainable; it does not 'really' exist for the physicist.

IV

THE MATHEMATICAL FORMULATION OF SPATIAL RELATIVITY

In mathematical phraseology we can express this result by saying: two worlds, which can be transformed into one another by a perfectly arbitrary (but continuous and one-to-one) point-transformation, are, with respect to their physical reality, *identical*. That is: if the universe is deformed in any way, so that the points of all physical bodies are displaced to new positions, then (taking account of the above supplementary considerations), no measurable, no 'real' change has happened at all, if the co-ordinates of a physical point in the new position are any arbitrary functions whatsoever of the co-ordinates of its old position. Of course, it will have to be postulated that the points of the bodies retain their connexion, and that points which were neighbouring before the deformation remain so after it (i.e. these functions must be continuous); and, moreover, to every point of the original world only *one* point of the new world must correspond, and vice versâ (i.e. these functions must be one-valued).

It is easy to picture the relations described by imagining space to be divided by three families of planes, respectively parallel to the co-ordinate planes, into a number of little cubes. Those points of the world, which lie on such a plane (e.g. the ceiling of a room) will, after the deformation, form a more or less bent surface. The second world will thus be divided by the system of these bent planes into eight-cornered cells, which will in general be different in size and form. But in this world, we should, just as before, denominate these surfaces 'planes' and their curves of intersection 'straight lines', and the cells 'cubes'; for every means of proving that they are not 'really' so would be lacking. If we suppose the planes numbered in order, then every physical point of the deformed world is defined by three numbers, namely the numbers of the three surfaces which intersect at it; we can thus use these numbers as co-ordinates of this point, and shall fittingly call them 'Gaussian co-ordinates', since they signify the same for three-dimensional configurations as the co-ordinates which Gauss in his time introduced for the examination of two-dimensional configurations (surfaces). He supposed two intersecting families of curves to be drawn on any arbitrarily curved surface in such a way as to lie entirely on the surface. Each surface-point is then defined by specifying the two curves (one member from each family) which pass through the point. It is now evident that with these assumptions the bounding surfaces of bodies, the path of light-rays, all motions and all natural laws in the deformed world, expressed in these new co-ordinates, will be represented by identically the same equations as the corresponding objects and events of the original world, referred to ordinary Cartesian co-ordinates, provided that the numbering of the surfaces is carried out correctly. A difference between the two worlds exists, as we have said, only so long as one erroneously supposes that planes and lines can be defined in space at all without reference to bodies in it, as if it were endowed with 'absolute' properties.

But, if we regard the *old* co-ordinates, i.e. the system of perpendicularly intersecting planes, from the point of view of the new universe, *these* planes will now—reciprocally—seem to be an entirely curved and distorted system; and geometrical forms and physical laws, when referred to this system, assume an entirely new appearance. Thus, instead of saying that I deform the world in a certain way, I can equally well say that I am describing the unchanged world by means of new co-ordinates, the plane-system of which is deformed in some definite way as compared with the first. Both processes are truly the same; and these imaginary deformations would not signify any real alteration of the world, but merely a reference to other co-ordinates.

We may therefore also regard the world in which we live as the distorted one, and say that the surfaces of bodies (e.g. the ceiling of a room), which we call planes, are not 'really' such; our straight lines (light-rays) are 'in reality' curved lines, &c. We could, without any contradiction manifesting itself, assume that a cube which is taken into another room alters its shape and size considerably on the way; we should not be aware of the change, because we ourselves, with all measuring instruments and the whole surroundings, suffer analogous changes; certain curved lines would have to be considered as the 'true' straight lines. The angles of our cubes, which we call right angles, would, 'in reality', not be so—yet we could not establish this: since the measure by means of which we have determined the arms of the angles would correspondingly change in length, when we turned it round to measure the circular arc belonging to the angle. The sum of the angles of our square would 'in reality' not amount to four right angles—in short, it would be as if we used a geometry other than Euclidean. The whole assumption would be tantamount to maintaining that certain surfaces and lines, that appear curved to us, are really 'true' planes and straight lines, and that we should have to use them as co-ordinates.

Why do we not actually suppose anything of the sort, although it would be theoretically possible, and although all our observations could be explained by this means? Simply because this explanation could be given only in a very complicated way, viz. by assuming extremely intricate physical laws. The shape of a body would be dependent upon its position; it would, if sufficiently far removed from the influence of external forces, describe a curved line, &c.; in a word, we should arrive at a very involved system of physics, and—most important of all—it would be quite arbitrary; for there would be an unlimited number of similarly complicated systems of physics, which would all serve equally well for describing Experience. Compared with these, the usual system, which applies Euclidean geometry, distinguishes itself as the *simplest,* as far as can be judged up to the present. The lines which we call 'straight' play a special rôle in physics; they are, as Poincaré expresses it, more 'important' than other lines. A co-ordinate system founded on these lines therefore leads to the simplest formulae for physical laws.

V

THE INSEPARABILITY OF GEOMETRY AND PHYSICS IN EXPERIENCE

THE reasons for preferring the usual system of geometry and physics to all other possible ones, and for considering it to be the only 'true' one, are exactly the same as those which make the Copernican view of the world superior to that of Ptolemy; the former leads to a much simpler system of celestial mechanics. The formulation of the laws of planetary motions become excessively complicated, if we refer them, as Ptolemy did, to a co-ordinate system rigidly attached to the earth; on the other hand, the process becomes quite simple, if a co-ordinate system which is at rest with respect to the fixed stars be chosen.

We thus see that experience in no wise compels us to make use of an absolute geometry, e.g. that of Euclid, for the physical description of nature. It teaches us only what geometry we must use, if we wish to arrive at the simplest formulae to express the laws of physics. From this it immediately follows that there is no meaning in talking of an absolute geometry of 'space', omitting all reference to physics and the behaviour of physical bodies; for, since experience leads us to choose only a certain geometry, in that it shows us in what way the behaviour of bodies can be described most simply in mathematical language, it is meaningless to attempt to assign a distinctive position to any one geometry, as long as we leave material bodies out of account. Poincaré has expressed this tersely in the words: 'Space

itself is amorphous; only the things in it give it a form.'
I shall just recall a few remarks of Helmholtz, in which
he expresses the same truth. At the conclusion of his
lecture on the *Origin and Significance of the Axioms of
Geometry,* he says: 'If for some particular reason we were
to find it expedient, we could quite logically consider the
space in which we live to be like the apparent space as
pictured in a convex mirror, wherein lines converge and the
background is contracted; or, we could take a limited spherical portion of our space, beyond the boundaries of which our
perceptions do not extend, and regard it as boundless
pseudo-spherical space. We should, in that case, have to
ascribe to bodies which appear rigid to us, and to our own
bodies at the same time, only the corresponding extensions
and contractions; and we should, of course, have to alter our
system of mechanical principles entirely. For even the
simple theorem that every point which is in motion and is not
acted on by any forces continues to move in a straight line
with invariable velocity, no longer holds true for the world
which is represented in a convex mirror. . . . Geometrical
axioms are in no way confined to relations in space alone,
but also make assertions about the mechanical behaviour of
our most rigid bodies when in motion.'

Since the time of Riemann and Helmholtz we have been accustomed to talk of plane, spherical, pseudo-spherical and
other spaces, and discriminate from our observations to
which of these classes our 'real' space belongs. We now understand how to interpret this: viz. *not* as if one of these can
be predicated of space, without taking account of objects in
it; but in the sense that experience teaches us only whether it
is more practical to use Euclidean or non-Euclidean geometry for the physical description of nature. Riemann himself, and likewise Helmholtz, was quite clear about the question; but the results of both these investigators have often

been misinterpreted, so that they have occasionally even been used to strengthen the belief that absolute space has a particular form of its own ascertainable from experience. We must be on our guard against assuming that space has any 'physical reality' in this sense. It is well known that Gauss tried to measure directly, by means of theodolites, whether the sum of the angles of a very large triangle amount to two right angles or not. That is, he measured the angles which three light-rays, emitted from three fixed points (The Brocken, Hoher Hagen, and Inselberg), made with each other. Supposing that a deviation from two right angles had manifested itself, we could *either* regard the light-rays as curved and still use Euclidean geometry, *or* we could still call the path of a light-ray straight, but we should then have to introduce a non-Euclidean geometry. It is therefore not correct to say that experience could ever *prove* space to be 'non-Euclidean in structure', i.e. could ever compel us to adopt the second of these alternatives. On the other hand, Poincaré also errs when he somewhere expresses the opinion that the physicist would actually always choose the first assumption. For no one was able to predict whether it might not some time be necessary to depart from Euclidean measure-determinations in order to be able to describe the physical behaviour of bodies most simply.

All that could be affirmed at that time was that we should never find occasion to depart from Euclidean geometry to any *considerable* degree, since otherwise our observations, particularly in astronomy, would long ago have called our attention to this fact. Hitherto, however, by using Euclidean geometry as a foundation, we have admirably succeeded in arriving at simple physical principles. From this we may conclude that it will always be suited for at least an approximate description of physical events. If, therefore, to attain simplicity of expression, it should prove convenient to give up Euclidean measure-determinations in physics, such resulting deviations could only be slight, and would show differences only in regions on the outskirts of our field of observation. The essential significance of these deviations, whether great or small, naturally remains the same.

This case, hitherto only a theoretical possibility, has now presented itself. Einstein shows that non-Euclidean relations must actually be used in representing spatial conditions in physics so that it may be possible to maintain the extraordinary simplification of the principles underlying our view of physical nature, as embodied in the *general* theory of relativity. We shall return to this point presently. Meanwhile, we shall accept the result that space itself in no wise has a form of its own; it is neither Euclidean nor non-Euclidean in constitution, just as it is not a peculiarity of distance to be measured in kilometres and not in miles. In the same way as a distance only acquires a definite length when we have chosen a particular measure as unit, and in addition set out the mode of measurement, so a definite geometry can be applied to physical reality only when a definite method has been fixed upon, according to which spatial conditions are to be abstracted from physical conditions. Every measurement of spatial distances, when reduced to the essentials, is performed by placing one body against another; if such a comparison between two bodies is to become a *measurement*, it must be *interpreted* by taking due account of certain principles (e.g. one must assume that certain bodies are to be regarded as rigid, i.e. endure a translation without change of form). Precisely similar reflections may be made *mutatis mutandis* for time. Experience cannot compel us to found our description of physical nature upon a definite measure and rate of time; we choose just that measure and rate which enable us to formulate physical laws most simply. All time-determinations are just as indissolubly associated with physical occurrences as spatial ones are with physical bodies. Quantitative observations of any physical occurrence, such as e.g. the propagation of light from one point to another, imply that readings must be taken from a clock, and thus assume a method according to which clocks in different localities are to be regulated with one another. Without this means, the conceptions of simultaneity and equal duration have no definite meaning. These are matters to which we called attention earlier, when we were discussing the special theory of relativity. All time-measurements are undertaken by comparing two events, and if they are to have the significance of a true measurement, some convention or principle must be assumed, the choice of which will again be determined by the endeavour to obtain physical laws in the simplest form.

We thus see: Time and Space can be dissociated from physical things and events only in abstraction, i.e. mentally. The combination or oneness of space, time, and things is alone reality; each by itself is an abstraction. Whenever we make an abstraction, we must always ask whether it has a physical meaning, i.e. whether the products of abstraction are actually independent of one another.

VI

THE RELATIVITY OF MOTIONS AND ITS CONNEXION WITH INERTIA AND GRAVITATION

IF one had not lost sight of this last truth, the celebrated controversy, whch was always being renewed, about so-called *absolute motion* would from the very outset have assumed a different aspect. The conception of motion has, in the first place, a real meaning only in dynamics, as the change of position of material bodies with time; so-called pure kinematics (known as 'phoronomy' in Kant's time) arises out of dynamics by abstracting from *mass*, and is thus the time-change of the position of mere mathematical points. How far this product of abstraction may serve for describing physical nature can be decided only by experience. Before the time of Einstein, the opponents of absolute motion (e.g. Mach) always argued thus: Every determination of position, being only defined for a definite system of reference, is, as regards its conception, relative, and therefore also every change of position. Hence only relative motion exists, i.e. there can be no unique system of reference; for, since the conception of rest is only relative, I must be able to regard every system of reference as being at rest. This method of proof, however, overlooks the fact that the definition of motion as being *merely change of position* applies to motion only in the kinematical sense. For real motions, i.e. for mechanics or dynamics, this conclusion need not be regarded as final; experience must prove whether it is justified. From the purely kinematical point

of view, it is, of course, the same to say that the earth
rotates as that the stellar heavens are rotating around
the earth. It does not follow, however, that both state-
ments are indistinguishable dynamically. Newton, as is
known, assumed the contrary. He believed—apparently in
perfect agreement with experience—that a rotating body
could be distinguished from one at rest by the appearance
of centrifugal forces (with resultant flattening); and absolute rest (leaving out of account any motion of uniform
translation) would be defined by the absence of centrifugal
forces. In realizable experience, every accelerated change
of position is accompanied by the appearance of inertial resistances (e.g. centrifugal forces); and it is quite arbitrary
to declare one of these factors, which both belong equally to
physical motion, and are only separable in abstraction, to be
the cause of the other, viz. to regard the inertial resistances
as the *effect* of the acceleration. It cannot therefore be
proved out of the mere conception of motion (as Mach endeavoured to do) that there can be no unique system of reference, i.e. that there can be no absolute motion; the decision can only be left to observation.

Newton certainly erred in believing that observation *had*
already decided this question, viz. in the sense that two uniform rectilinear motions were in fact relative (i.e. that the
laws of dynamics are exactly the same for two systems of
reference which are moving uniformly and rectilinearly
with regard to one another), but that this was not true for
accelerated motions (e.g. rotations). Accelerations, he
thought, were of an absolute nature; certain systems of reference were unique in that the Law of Inertia held for them
alone. They were therefore called Inertial Systems. According to Newton, an Inertial System would thus be defined and recognizable as one in which a body, upon which
no forces act, would move uniformly and rectilinearly (or

remain at rest); and consequently centrifugal forces (or flattening) would only fail to manifest themselves in or on a body if it were not rotating with reference to the inertial system. Newton used these views as a foundation for mechanics unjustifiably; for actually they are *not* sufficiently founded on experience. No observation shows us a body on which no forces are acting,[1] and no experience has yet proved whether a body which is at rest in an inertial system might not be subject to centrifugal forces if an extraordinarily great mass were to rotate near it, i.e. whether these forces are not, after all, only peculiarities of *relative* rotation.

The state of affairs was in fact as follows. On the one hand, the experiences so far known did not suffice to prove the correctness of Newton's assumption that absolute accelerations existed (i.e. unique systems of reference); on the other hand, the general arguments in favour of the relativity of all accelerations, e.g. Mach's, were not, as we have just shown, conclusive. From the standpoint of actual experience, both points of view had for the time being to be considered admissible. But, regarded philosophically, the standpoint which denied the existence of unique systems of reference, thus affirming *all* motions to be relative, is very attractive, and possesses great advantages over the Newtonian view; for, if it were realizable, it would signify an extraordinary simplification of our picture of the world. It would be exceedingly satisfactory to be able to say that not only uniform, but indeed all, motions are relative. The kinematical and dynamical conception of motion would then become identical in essence. To determine the character of motion, purely kinematical observations would suffice. It would not be necessary to add observations about centrifu-

[1] Mach and Pearson called particular attention to this. Karl Pearson, *Grammar of Science*, Chap. VIII, § 4.

gal forces, as it was for the Newtonian view. A system of mechanics built up on relative motions would thus result in a much more compact and complete view of the world than that of Newton. It would not indeed (as was apparently the opinion of Mach) be proved to be the only *correct* view of the universe; but (as Einstein points out) it would recommend itself from the very outset by its imposing simplicity and finish.[1]

Up to the time of Einstein, however, such a world-view, i.e. the idea of a system of mechanics founded on relative motions, had been only a desire, an alluring goal; such a system of mechanics had never been enunciated, nor had a possible way to it even been pointed out. There was no means of knowing whether, and under what conditions, it was possible at all or compatible with empirical facts. Indeed, science seemed to be constrained to develop in the contrary direction; for, whereas in classical mechanics all systems moving uniformly and *rectilinearly* with respect to one inertial system were likewise inertial systems (so that at least all uniform motions of translation preserved

---

[1] Einstein adds that Newton's mechanics only seemingly satisfies the demands of causality, e.g. in the case of bodies which are rotating and suffer a flattening. But this mode of expression does not appear to me to be quite free from objection. We need not look upon the Newtonian doctrine as making Galilean space, which is of course not an observable thing, the *cause* of centrifugal forces; but we can also consider the expression 'absolute space' to be a paraphrase of the mere fact that these forces exist. They would then simply be immediate data; and the question why they arise in certain bodies and are wanting in others would be on the same level with the question why a body is present at one place in the world and not at another. Absolute rotation need not be regarded as the *cause* of the flattening, but we can say that the former is *defined* by the latter. In this way I believe that Newton's dynamics is quite in order as regards the principle of causality. It would be easy to defend it against the objection that purely fictitious causes are introduced into it, although Newton's own formulation was incorrect.

the character of being relative), in the case of electromagnetic and optical phenomena even this no longer seemed to hold; in Lorentz's Electrodynamics there was only one unique system of reference (the one which is ' at rest in the aether '). Only after Einstein had succeeded in extending the special principle of relativity, which was valid in classical mechanics, to *all* physical phenomena, could the idea of the entirely general relativity of any arbitrary motions again be taken up on the ground thus prepared; and again it was in the hands of Einstein that it bore fruit. He transplanted it as it were from regions of philosophy to those of physics, and thereby brought it within the range of scientific research.

Although the philosophical arguments were so powerful in themselves, Einstein gave them additional weight by adding to them the physical argument that all motions were most probably endowed with a relative character. This physical argument is built on the equality of inertial and gravitational mass. We can see it more clearly in the following way. If we assume all accelerations to be relative, then all centrifugal forces, or other inertial resistances which we observe, must depend on motion relative to other bodies; we must therefore seek the cause of these inertial resistances in the presence of those other bodies. If, for example, there were no other body present in the heavens except the earth, we could not speak of a rotation of the earth, and the earth could not be flattened at the poles. The centrifugal forces, as a consequence of which the earth's flattening comes about, must thus *owe* their existence to the action on the earth of the heavenly bodies. Now, as a matter of fact, classical mechanics is acquainted with an action which all bodies exert on one another, viz. *Gravitation.* Does experience lend any support to the suggestion that this gravitational influence might be made answerable for the inertial effects? This support is actually to be found, and is very remarkable; it consists in the circumstance that one and the same constant plays the determining rôle for both inertial and gravitational effects, viz. the quantity known as *mass*. If, for instance, a body describes a circular path relatively to an inertial system, the necessary central force is, according to classical mechanics, proportional to a factor *m* which is a characteristic for the body; but if the body is attracted by another body (e.g. the earth) in virtue of gravitation, the force acting on it (e.g. its weight) is proportional to this same factor *m*. It is on account of this that, at the same place in a gravitational field, all bodies without exception suffer the *same* acceleration; for the mass of a body eliminates itself, since it occurs as a factor of proportionality both in the expression for the inertial resistance and in that for the attraction.

Einstein has made the connexion between gravitation and inertia extraordinarily clear by the following reflection. If a physicist, enclosed in a box somewhere out in space, were to observe that all objects left to themselves in the box acquired a certain acceleration, e.g. fell to the bottom with constant acceleration, he could interpret this phenomenon in two ways:—in the first place, he could assume that his box was resting on the surface of some heavenly body, and he would then ascribe the falling of the objects to the gravitational influence of the heavenly body; or, he could assume instead that the box was moving 'upwards' with constant acceleration, and then the behaviour of the 'falling' bodies would be explained by their inertia. Both explanations are equally possible, and the enclosed physicist would have no means of discriminating between them. If we now assume that all accelerations are relative, and that a means of discrimination is *essentially* wanting, this may be generalized. We may consider the observed acceleration of any body left to itself, at any point in the universe, to be due to the effect either of inertia or of gravitation, i.e. we may either say 'the system of reference, from which I am observing this event, is accelerated' or 'the event is taking place in a gravitational field'. We shall follow Einstein, and call the statement that both interpretations are equally justifiable the *Principle of Equivalence*. It is founded, as we have seen, on the identity of inertial and gravitational mass.

The circumstance of the identity of these two factors is very striking, and when we get to realize its full import, it seems astonishing that it did not occur to any one before Einstein to bring gravitation and inertia into closer connexion with one another. If something analogous had been observed in another branch of physics (e.g. if an effect had been found which was proportional to the quantity of electricity associated with a body) we should immediately have brought it into relationship with the remaining electrical phenomena; we should have regarded electrical forces, and the supposed new effect, as different manifestations of one and the same governing principle. In classical mechanics, however, not the slightest connexion was introduced between gravitational and inertial phenomena; they were not comprised under one sole principle, but existed side by side totally unrelated. The fact that one and the same factor— mass—played a similar part in each seemed mere chance to Newton. Is it really only chance? This seems improbable in the highest degree.

The identity of inertial and gravitational mass is thus the real ground of experience which gives us the right to assume or assert that the inertial effects which we observe in bodies are to be traced back to the influence which is exerted upon them by other bodies. (This influence is, of course, in accordance with modern views, to be conceived not as an action at a distance, but as being transmitted through a field.)

The above assertion (of identity) implies the postulate of an unlimited relativity of motions; for, since all phenomena are to depend only on the *mutual* position and motion of bodies, reference to any particular co-ordinate system no longer occurs. The expression of physical laws, with reference to a co-ordinate system attached to any arbitrary body (e.g. the sun), must be the same as with reference to one attached to any other body whatsoever (e.g. a merry-go-round on the earth); we should be able to look upon both with equal right as being 'at rest'. The laws of Newtonian mechanics had to be referred to a perfectly definite system (an Inertial System) which was quite independent of the mutual position of bodies; for the Law of Inertia held for these only. In the new mechanics, on the other hand, which has to look upon inertial and gravitational forces as the expression of a single fundamental law, not only gravitational phenomena, but also inertial phenomena, are to depend exclusively on the position and motion of bodies relative to one another. The expression for this fundamental law must accordingly be such that no co-ordinate system plays a unique part compared with the others, but that all remain valid for any arbitrary system. It is evident that the old Newtonian dynamics can signify only a first approximation to the new mechanics; for the latter demands, in contradistinction to the former, that centrifugal accelerations, for example, must be induced in a body if large masses rotate around it; and the contradiction between the new theory and classical mechanics does not come into evidence in this particular case, merely because these forces are so small, even for the greatest available masses in the experiment, that they escape our observation.

Einstein has actually succeeded in establishing a fundamental law which comprises inertial and gravitational phenomena alike. We are now better prepared to follow the line of argument by which Einstein arrived at this result.

VII

THE GENERAL POSTULATE OF RELATIVITY AND THE MEASURE-DETERMINATIONS OF THE SPACE-TIME CONTINUUM

THE idea of relativity has only been applied in the preceding pages to physical thought in so far as it bears on motions. If these are really relative without exception, any coordinate systems moving arbitrarily with reference to one another are equivalent, and space loses its objectivity, in so far as it is not possible to define any motions or accelerations with respect to it. Yet it still preserves a certain objectivity, so long as we tacitly imagine it to be provided with absolutely definite metrical properties. In the older physics every process of measurement was unhesitatingly founded on the notion of a rigid rod, which preserved the same length at all times, no matter what its position and surroundings might be; and proceeding from this, all measurements were determined according to the rules of Euclidean geometry. This process was not changed in any way in the new physics which is based on the special theory of relativity, provided that the condition was fulfilled that the measurements were all carried out within the same co-ordinate system, by means of a rod respectively at rest with regard to each system in question. In this way space was still endowed with the independent property, as it were, of being 'Euclidean' in 'structure', since the results of these measure-determiations were regarded as being entirely independent of the physical conditions prevailing in space, e.g. of the distribution of bodies and their gravitational fields. Now we have seen that it is always possible to fix the position- and magnitude-relations of bodies and events according to the ordinary Euclidean rule, e.g. by means of Cartesian co-ordinates, so long as the laws of physics have been correspondingly formulated. But we are subject to a limitation: we had set out to determine them, if possible, in such a manner that the general postulate of relativity would be fulfilled. Now it by no means follows that we shall succeed *in fulfilling this condition* if we use Euclidean geometry. We have to take into account the possibility that this may not be so. Just in the same way as we found that the postulate of special relativity could be satisfied only if the conception of time which had previously prevailed in physics was modified, it is likewise quite possible that the generalized principle of relativity might compel us to depart from ordinary Euclidean geometry.

Einstein, by considering a very simple example, comes to the conclusion that we are actually compelled to make this departure. If we fix our attention upon two rotating co-ordinate systems, and assume that in one of them, say K, the positional relations of the bodies at rest (in K) can be determined by means of Euclidean geometry (at least in a certain domain of K), then this is certainly not possible for the second system K^1. This is easily seen as follows. Let the origin of co-ordinates and the z-axis of the two systems coincide, and let the one system rotate relatively to the other about this common axis. We shall suppose a circle described about the origin as centre in the x-y-plane of K; for reasons of symmetry this is also a circle in K^1. If Euclidean geometry holds in K, then the ratio of the circumference to the diameter is in this system π; but if we determine this same ratio by means of measurements with

rods which are at rest in K^1, we obtain a value greater than
π. For, if we regard this process of measurement from the
system K, the measuring-rod has the same length in measuring the diameter as if it were at rest in K: whereas in
measuring the circumference it is shortened, owing to the
Lorentz-Fitzgerald contraction; the ratio of these numbers
thus becomes greater than π and the geometry which holds
in K^1 is *not* Euclidean. Now, the centrifugal forces with
respect to K^1, which are due to inertial effects (on the old
theory), may, however, be regarded at every point, according to the Principle of Equivalence, as gravitational effects.
From this it can be seen that the existence of a gravitational field demands that non-Euclidean measure-determinations be used. Strictly speaking, there is, however, no finite
domain which is entirely free from gravitational effects; so
that, if we wish to maintain the postulate of general relativity, we must refrain from describing metrical and positional relations of bodies by Euclidean methods. This does
not mean that in place of Euclidean geometry we are now to
use some other definite geometry, such as that of Lobats-
chewsky or Riemann, for the whole of space (cf. Section IX
below), but that all types of measure-determination are to
be used: in general, a different sort at every place. Which
it is to be, depends upon the gravitational field at the place.
There is not the slightest difficulty in thinking of space in
this way; for we fully convinced ourselves above that it is
only the things in space which give it a definite structure
or constitution; and now we have only to assign this rôle—
as we shall immediately see—to gravitational masses or
their gravitational fields respectively. It becomes impossible
to define and measure lengths and times (as may likewise
easily be shown) in a gravitational field in the simple manner described in Section II, by means of clocks and measuring-rods. Since gravitational fields are nowhere absent, the

special theory of relativity nowhere holds accurately; the velocity of light, for instance, is never in truth absolutely constant. It would, however, be quite wrong to say that the special theory had been proved to be false, and had been overthrown by the general theory. It has really only been assimilated in the latter. It represents the special case into which the general theory resolves when gravitational effects become negligible.

It follows, then, from the general theory of relativity that it is quite impossible to ascribe any properties to space without taking into account the things in it. The relativization of space has thus been carried out completely in physics, as was shown by the above general considerations to be the most likely result. Space and Time are never objects of measurement in themselves; only conjointly do they constitute a four-dimensional scheme, into which we arrange physical objects and processes by the aid of our observations and measurements. We choose this scheme in such a way that the resultant system of physics assumes as simple a form as possible. (We are free to choose, since we are dealing with a product of abstraction.)

How is this arrangement to be fitted into the scheme? What is it that we really observe and measure?

It is easily seen that the possibility of observing accurately depends upon noting identically the same physical points at various times and in various places; and that all measuring reduces itself to establishing that two such points, upon which we have fixed, coincide at the same place and at the same time. A length is measured by applying a unit measure to a body, and observing the coincidence of its ends with definite points on the body. With our apparatus the measurement of all physical quantities resolves finally into the measurement of a length. The adjustment and reading of all measuring instruments of whatsoever variety—whether they be provided with pointers or scales, angular-diversions, water-levels, mercury columns, or any other means—are always accomplished by observing the space-time-coincidence of two or more points. This is also true above all of apparatus used to measure time, familiarly termed *clocks*. Such coincidences are therefore, strictly speaking, alone capable of being observed; and the whole of physics may be regarded as a quintessence of laws, according to which the occurrence of these space-time-coincidences takes place. Everything else in our world-picture which can *not* be reduced to such coincidences is devoid of physical objectivity, and may just as well be replaced by something else. All world pictures which lead to the same laws for these point-coincidences are, from the point of view of physics, in every way equivalent. We saw earlier that it signifies no observable, physically real, change at all, if we imagine the whole world deformed in any arbitrary manner, provided that *after* the deformation the co-ordinates of every physical point are continuous, single-valued, but otherwise quite arbitrary, functions of its co-ordinates *before* the deformation. Now, such a point-transformation actually leaves all spatial coincidences totally unaffected; they are not changed by the distortion, however much all distances and positions may be altered by them. For, if two points A and B, which coincide before the deformation (i.e. are infinitely near one another), are at a point the co-ordinates of which are x_1, x_2, x_3, and if A arrives at the point x_1', x_2', x_3', as a result of the deformation, then, since by hypothesis the x''s are continuous single-valued functions of the x's, B must also have the co-ordinates x_1', x_2', x_3', after the deformation—i.e. must be at the same point (or infinitely near) A. Consequently, all coincidences remain undisturbed by the deformation.

Earlier, we had only, for the sake of clearness, investi-

gated these effects in the case of space; we may now generalize by adding the time t as a fourth co-ordinate. Better
still, we may choose as our fourth co-ordinate the product ct
$(= x_4)$ in which c denotes the velocity of light. These are
conventions which simplify the mathematical formulation
and our calculations, and have a merely formal significance
for the present. It would therefore be wrong to associate
any metaphysical speculations with the introduction of the
four-dimensional point of view.

Over and above its convenience for this formulation, we
can see other advantages which accrue from our regarding
time as a fourth co-ordinate, and recognize therein an essential justification for this mathematical view. To show this
clearly, let us suppose a point to move in any way in a plane
(that of x_1-x_2 may be chosen). It describes some curve in
this plane. If we draw this curve, we can, by looking at
it, get an impression of the shape of its path, but not of any
other data of its motion, e.g. the velocity which it has at
different points of its path, or the time at which it passes
through these points. But if we add time x_4 as a third co-
ordinate, the same motion will be represented by a three-
dimensional curve, the form of which immediately gives us
information about the character of the motion; for we can
recognize directly from it which x_4 belongs to any point x_1 x_2
of the path, and we can also read off the velocity at any moment from the inclination of the curve to the x_1-x_2-plane.
We shall follow Minkowski by appropriately calling this
curve the *world-line* of the point. A circular motion in the
x_1-x_2-plane would be represented by a helical world-line in
the x_1-x_2-x_4-manifold. This trajectory of the point only
arbitrarily expresses, as it were, one aspect of its motion,
viz. the projection of the three-dimensional world-line on the
x_1-x_2-plane. Now, if the motion of the point itself takes
place in three-dimensional space, we obtain for its world-

line a curve in the four-dimensional manifold of the x_1, x_2, x_3, x_4, and from this line all characteristics of the motion of the point can be studied with the greatest ease. The path of the point in space is the projection of the world-line on the manifold of the x_1, x_2, x_3, and thus gives an arbitrary and one-sided view of a few properties only of the motion: whereas the world-line expresses them all in their entirety.

Our considerations about the general relativity of space may immediately be extended to the four-dimensional space-time manifold; they apply here also, for to increase the number of co-ordinates by one does not alter the underlying principle. The system of world-lines in this x_1-x_2-x_3-x_4-manifold represents the happening in time of all events in the world. Whereas a point transformation *in space alone* represented a deformation of the world, i.e. a change of position and a distortion of bodies, a point-transformation in the four-dimensional universe also signifies a change in the state of *motion* of the three-dimensional world of bodies: since the time co-ordinate is also affected by the transformation. We can always imagine the results which arise from the four-dimensional forms, by picturing them as motions of three-dimensional configurations. If we suppose a complete change of this sort to take place, by which every physical point is transferred to another space-time point in such a way that its new co-ordinates, x'_1, x'_2, x'_3, x'_4, are quite arbitrary (but continuous and single-valued) functions of its previous co-ordinates x_1, x_2, x_3, x_4: then the new world is, as in previous cases, not in the slightest degree different from the old one physically, and the whole change is only a transformation to other co-ordinates. For that which we can alone observe by means of our instruments, viz. space-time-coincidences, remains unaltered. Hence points which coincided at the world-point x_1, x_2, x_3, x_4 in the one universe would again coincide in the other at the world-point x'_1, x'_2, x'_3, x'_4. Their coincidence—and this is all that we can observe—takes place in the second world precisely as in the first.

The desire to include, in our expression for physical laws, only what we physically observe leads to the postulate that the equations of physics do not alter their form in the above arbitrary transformation, i.e. that they are valid for any space-time co-ordinate systems *whatever*. In short, expressed mathematically, they are 'covariant' for *all* substitutions. This postulate contains our general postulate of relativity; for, of course, the term '*all* substitutions' includes those which represent transformations of entirely arbitrary three-dimensional systems in motion. But it goes further than this, inasmuch as it allows the relativity of space, in the most general sense discussed above, to be valid even *within* these co-ordinate systems. In this way Space and Time are deprived of the 'last vestige of physical objectivity', to use Einstein's words.

As explained above,[1] we may determine the position of a point by supposing three families of surfaces to be drawn through space, and then, after assigning a definite number, a parametric value, to each successive surface of each family, we may regard the numbers of the three surfaces which intersect at the point as its co-ordinates. (Each family must be numbered independently of the others.) Of course, the relations between co-ordinates which are defined in this way (Gaussian co-ordinates) will not in general be the same as those which hold between the ordinary Cartesian co-ordinates of Euclidean geometry. The Cartesian x-co-ordinate of a point, for example, is ascertained by marking off the distance from the beginning of the x-axis by means of a rigid unit measure; the number of times this measure has to be

[1] Page 29.

applied end to end gives the desired co-ordinate number. In the case of the new co-ordinates other conditions hold (cf. page 48 above), since the value of a parameter is not immediately obtainable as a number by applying the unit measure. We must consequently regard the x_1, x_2, x_3, x_4 of the four-dimensional world as parameters, each of which represents a family of three-dimensional manifolds; the space-time continuum is partitioned by four such families, and four three-dimensional continua intersect at each world-point, their parameters thus being its co-ordinates.

If we now consider that the principle by which the co-ordinates are to be fixed consists in a perfectly arbitrary partition of the continuum by means of families of surfaces —for, physical laws are to remain invariant for *arbitrary* transformation—it seems at first sight as if we no longer had any firm footing or means of orientation. We do not immediately see how measurements are possible at all, and how we can succeed in ascribing definite number values to the new co-ordinates, even if these are no longer directly results of measurement. Comparing measuring-rods and observing coincidences result in a *measurement,* as we have seen, only if they are founded on some idea, or some physical assumption or, rather, convention; the choice of which, strictly speaking, is essentially of an arbitrary nature, even if experience points so unmistakably to it as being the simplest that we do not waver in our selection. We therefore find it necessary to make some convention, and we arrive at this by a sort of principle of continuation, as follows. In ordinary physics we are accustomed to assume without argument that we may speak of rigid systems of reference, and can realize them to a certain degree of approximation; length may then be regarded as being one and the same quantity at every arbitrary point, in every position and state of motion. This assumption had already been modified to a certain extent in the special theory of relativity. According to the latter, the length of a rod is in general dependent upon its velocity relative to the observer; and the same holds of the indications of a clock. The connexion with the older physics, and, as it were, the continuous transition to it, are due to the circumstance that the alterations in the length- and time-data become imperceptibly small, if the velocity is not great; for small speeds (compared with those of light) we may regard the assumptions of the old theory as being allowable. The special theory of relativity so adjusts its equations that they degenerate into the equations of ordinary physics for small velocities.

In the general theory, the relativity of lengths and time goes much further still; the length of a rod, according to it, can also depend on its place and its position. To gain a starting point at all, a Δός μοι ποῦ στῶ, we shall of course maintain continuity with the physics which has hitherto proved its worth, and accordingly assume that this relativity vanishes for extremely small changes. We shall thus consider the length of a rod to remain constant as long as its place, its position, and its velocity change only slightly—in other words, we shall adopt the convention that, for infinitely small domains, and for systems of reference, in which the bodies under consideration possess no acceleration, the special theory of relativity holds. Since the special theory uses Euclidean measure-determinations, this includes the assumption that, for the systems designated above, Euclidean geometry is to remain valid for infinitely small portions. (Such an infinitely small domain may still be large compared with the dimensions which are used elsewhere in physics.) The equations of the general theory of relativity must be, in the special case mentioned, transformed into those of the special theory. We have now founded our theory on an idea which makes measurement possible, and we have reviewed the assumptions by means of which we can successfully solve the problem proposed by the postulate of general relativity.

VIII

ENUNCIATION AND SIGNIFICANCE OF THE FUNDAMENTAL LAW OF THE NEW THEORY

In accordance with the last remarks, we shall turn our attention to the realm of the infinitely small, and in it choose a three-dimensional Euclidean system of co-ordinates, in such a way that the bodies which are to be considered have no perceptible acceleration with respect to it. This choice is equivalent to the introduction of a definite four-dimensional co-ordinate system for the domain in question. Let us fix any point-event in this domain, i.e. a world-point A in the space-time-continuum, the co-ordinates of which we shall assume to be X_1, X_2, X_3, X_4, in our local system; of these X_1, X_2, X_3 are measured by applying a small measuring rule of unit length end to end, and the value of X_4 is determined by the reading of a clock. B is to represent a space-time point-event infinitely near A; its co-ordinates differ, by the values dX_1, dX_2, dX_3, dX_4, from those of A. The 'distance' of these two world-points is then given by the well-known simple formula

$$ds^2 = dX_1^2 + dX_2^2 + dX_3^2 - dX_4^2$$

This 'distance', the line-element of the world-line, connecting A and B, is, of course, not in general a space-distance [length], but, since it is a combination of space- and time-quantities, has the physical significance of a motional event, as we clearly pointed out in introducing the notion of world-lines. The numerical value of ds is always the same, whatever orientation the chosen local co-ordinate system may have.

(The special theory of relativity throws a clearer light on ds. If, for example, ds^2 is negative, it states that we can, by appropriately choosing co-ordinate directions, obtain $ds^2 =$ $-dX_4^2$, whilst the other three dX's vanish. There is then no difference between the space co-ordinates of the two world-points; the events corresponding to them thus occur in this system at the same place, but with a time-difference dX_4. In this case ds is said to belong to the 'time-class' of events; on the other hand it is assigned to the 'space-class' of events if 'ds^2' is positive; for in the latter case the co-ordinate directions may be so chosen that dX_4 vanishes. The two point-events then take place simultaneously for this system, and ds gives a measure of the distance which separates them. Finally $ds = 0$ signifies a motion which takes place with the velocity of light, as is easily seen if we substitute for dX_4 its value $c \cdot dt$.)

We shall now introduce any new co-ordinates x_1, x_2, x_3, x_4, which are quite arbitrary functions of X_1, X_2, X_3, X_4, i.e. we shall pass from our local system to any other arbitrary system. Certain co-ordinate differences dx_1, dx_2, dx_3, dx_4, correspond to the 'distancer' between the points A and B in this new system, and the old co-ordinate difference dX can be expressed in terms of the new dx's by using elementary formulae of the differential calculus.[1] If we insert the expressions thus obtained for the dX's in the above formula for the line-element, we obtain its value expressed in the new co-ordinates in the following form:

$$ds^2 = g_{11}dx_1^2 + g_{22}dx_2^2 + g_{33}dx_3^2 + g_{44}dx_4^2 + 2g_{12}dx_1dx_2 + 2g_{13}dx_1dx_3 + \ldots ,$$

[1] Viz.
$$dX_1 = \frac{\partial X_1}{\partial x_1}dx_1 + \frac{\partial X_1}{\partial x_2}dx_2 + \frac{\partial X_1}{\partial x_3}dx_3 + \frac{\partial X_1}{\partial x_4}dx_4,$$
$$dX_2 = \frac{\partial X_2}{\partial x_1}dx_1 + \frac{\partial X_2}{\partial x_2}dx_2 + \frac{\partial X_2}{\partial x_3}dx_3 + \frac{\partial X_2}{\partial x_4}dx_4, \&c.$$

i.e. as a sum of ten terms, in which the ten quantities g are certain functions of the co-ordinates X.[1] They do not depend on the particular choice of the local system, for the value of ds^2 was itself independent thereof.

When Riemann and Helmholtz examined three-dimensional non-Euclidean continua, they regarded the factors g, which occur above in the expression for the line-element, as purely geometrical quantities, by which the metrical properties of space were determined. They were perfectly aware, however, that we could not well speak of measurements and space without making some physical assumptions. Helmholtz's words were quoted above; here we need only allude to Riemann's remarks at the close of his inaugural dissertation (p. 268 of his Gesammelte Werke). He there states that, in the case of a continuous manifold, the principle of its measure-relations is not already contained in the conception of the manifold, but must 'come from elsewhere'; it is to be sought in 'binding forces', i.e. the ground of these measure-relations must be physical in nature. We know that reflections in the realm of metrical geometry acquire a meaning only when its relationship to physics is borne in mind. The above g's do not therefore merely allow a physical interpretation, but indeed demand it. Einstein's general theory of relativity gives them such an interpretation directly. For, to recognize the significance of the g's, we need only call to mind the physical meaning of the transformation from a local system to the general system, as was discussed just above. The former was de-

[1] By performing the operations indicated we easily find that:

$$g_{11} = \left(\frac{\partial X_1}{\partial x_1}\right)^2 + \left(\frac{\partial X_2}{\partial x_1}\right)^2 + \left(\frac{\partial X_3}{\partial x_1}\right)^2 - \left(\frac{\partial X_4}{\partial x_1}\right)^2 \&c.$$

$$g_{12} = \frac{\partial X_1}{\partial x_1}\frac{\partial X_1}{\partial x_2} + \frac{\partial X_2}{\partial x_1}\frac{\partial X_2}{\partial x_2} + \frac{\partial X_3}{\partial x_1}\frac{\partial X_3}{\partial x_2} - \frac{\partial X_4}{\partial x_1}\frac{\partial X_4}{\partial x_2}, \&c.$$

fined by the property that a material point, left to itself in the space of the X_1, X_2, X_3, moves rectilinearly and uniformly in this space; its world-line,[1] i.e. the law of its motion, is consequently a four-dimensional straight line, the line element of which is given by:

$$ds^2 = dX^2 + dX_2^2 + dX_3^2 - dX_4^2.$$

If we transform to the new co-ordinates x_1, x_2, x_3, x_4, this means that we are viewing the same event, the same motion of the point, from some other arbitrary system, with respect to which the local system is of course moving with acceleration in some way. Therefore, in the space of the x_1, x_2, x_3, the point moves curvilinearly and non-uniformly. The equation of its world-line, i.e. its law of motion, alters, inasmuch as its line-element, expressed in the new co-ordinates, is now given by:

$$ds^2 = g_{11} \ dx_1^2 + \ldots + g_{12} \ dx_1 \ dx_2 + \ldots$$

We now recall the 'Principle of Equivalence' (p. 41). According to this, the statement that 'a point left to itself moves with certain accelerations' is identical with the statement that 'the point is in motion under the influence of a gravitational field'. The equation of the world-line expressed in the new co-ordinates thus represents the motion of a point in the gravitational field. The factors g are hence the quantities which determine the field. We see that their part in the new theory is analogous to that played by the gravitational potential in the Newtonian theory. We may, therefore, term them the 10 components of the gravitational potential.

The world-line of the point, which was a straight line for the local system, i.e. the shortest connecting line between

---

[1] Its equation, expressed in the form of the shortest (geodetic) line, is

$$\delta \ (\textstyle\int ds) = 0.$$

two world-points, likewise represents a shortest line in the new system of x_1, x_2, x_3, x_4, for the definition of a geodetic line is independent of the co-ordinate system. If we could now regard the domains of the 'local' system as being infinitesimal, the whole world-line in it would shrink to an element ds. The reflection made above would become meaningless, and we could draw no further inferences. Since the Law of Inertia and the Special Theory of Relativity have, however, been so widely confirmed by experience, it is clear that there must in reality be finite regions, for which, if we choose a suitable system of reference, $ds^2 = dx_1{}^2 + dx_2{}^2 + dx_3{}^2 - dx_4{}^2$: viz. those parts of the world in which, with this chosen system, no perceptible influence of gravitating matter exists. In it the world-line is for this system a straight line, and consequently for arbitrary systems a geodetic line. We now again recall our Principle of Continuity (according to which the new laws are to be assumed, in such a way that the old laws are contained in them unchanged as nearly as possible, and the new ones resolve into the latter for the limiting case); and we then make the hypothesis that the relation obtained in this way is valid quite generally for *every* motion of a point under the influence of inertia and gravitation, i.e. that the world-line of the point is always a geodetic even when matter is present. This gives us the desired fundamental law. Whereas the Law of Inertia of Newton and Galilei states: 'A point under no forces moves uniformly and rectilinearly', the Einstein Law, which comprises both inertial and gravitational effects, asserts: *The world-line of a material point is a geodetic line in the space-time continuum.* This laws fulfils the condition of relativity; for it is an invariant for any arbitrary transformations, since the geodetic line is defined independently of the system of reference.

We must again emphasize that the co-ordinates $v_1 \ldots x_4$

are number-values, which fix the time and place of an event, but have not the significance of distances and times as measured in the ordinary way. The 'line-element' ds, on the other hand, has a direct physical meaning, and can be ascertained by means of measuring-scales and clocks. It is, by definition, independent of the system of co-ordinates; hence we need only betake ourselves to the local system of X_1 . . X_4, and the value which we there obtain for ds is valid generally.

Those steps have now been taken which are of general philosophic importance, and fundamental for the view of space and time according to the new doctrine: it is in these that we are here primarily interested. For Einstein they were merely the preliminary stage for the physical problem of getting at the actual values of the quantities g, i.e. of discovering how they depend upon the distribution and motion of the gravitating masses. In accordance with the Principle of Continuity, Einstein starts here again by working from the results of the special theory of relativity. The latter had taught us that not only matter in the ordinary sense, but also every kind of energy, has gravitational mass, and that inertial mass is altogether identical with energy. This implies that not the 'masses' but the energies [1] should figure in the differential equations giving the g's. The equations must of course remain covariant for any arbitrary substitutions. In addition to these initial assumptions which, from the point of view of the theory, are quite obvious, Einstein makes the further assumption that the differential equations are of the second order; he was guided by the fact that the old Newtonian potential satisfied a differential equation of just this type. In this way we arrive at perfectly definite equations for the g's, and

---

[1] They are represented in the special theory of relativity by the components of a four-dimensional 'tensor', the Impulse-energy tensor.

thus the problem of establishing them is (theoretically) solved.

So we see that, except for the last-mentioned purely formal analogy, the entire theory is built on foundations which have absolutely nothing in common with Newton's old theory of potential; it is, on the contrary, developed purely from the postulate of general relativity, and from well-known results of physics (as given by the special principle of relativity). It is so much the more surprising that the new equations, which have been obtained by such different means, actually degenerate into the Newtonian formula for general mass-attraction for a first approximation. This is in itself such an excellent confirmation of the lines of argument that it must inspire very considerable confidence in their correctness. But, as we know, the achievements of the new theory do not end here. For, if we work out the equations to a second approximation, there immediately emerges, without the help of any auxiliary assumptions, a quantitatively exact explanation of the anomalous motion of Mercury's perihelion, a phenomenon which the Newtonian Theory could account for only by introducing special hypotheses of a rather arbitrary nature. These are astonishing results, the scope of which cannot easily be overestimated: and we must agree with Einstein when he says at the conclusion of § 14 of his essay *Die Grundlage der allgemeinen Relativitätstheorie*': 'The fact that the equations deduced from the postulate of general relativity by purely mathematical processes . . . give us to a first approximation the Newtonian law of attraction, and to a second approximation the motion of Mercury's perihelion . . . discovered by Leverrier, is a convincing proof that the theory is physically correct'. The new fundamental law has an additional advantage over the Newtonian attraction formula, inasmuch as it is expressed as a differential law;

i.e. according to it, events at one point in the space-time manifold depend only upon the events of points infinitely near it on all sides, whereas in Newton's attraction formula gravitation occurs as a force acting at a distance. This means that we have considerably simplified the physical picture of the world, and consequently have now advanced another step in the theory of knowledge, by banishing gravitation, the last force acting at a distance, out of physics, and expressing all the laws underlying physical events solely by differential equations.

All the other laws must, of course, also be formulated in such a way that they remain invariant after any arbitrary transformations. The method of doing this is prescribed by the special principle of relativity and the principle of continuity, and has already been applied by Einstein and others. Chief interest circles around electrodynamics, from which it is to be hoped that, by combining it with the new theory of gravitation, it will be possible to build up a flawless system of physics. It is the great problem for physicists of the future to bring electrodynamics and gravitational theory under a common law, and thus embrace both realms in one theory. The endeavours which have been carried out in this direction have so far been unavailing; probably this is due, above all, to the absence of further data of experience, in which gravitational and electrical phenomena occur simultaneously.

In addition to the astronomical confirmation mentioned above, there are still other possibilities of verifying the theory by observation; for, according to it, there should be a still perceptible lengthening of the time of oscillation of light in a very strong gravitational field, and a curvature of the light rays should manifest itself. (The path of the latter being the geodetic lines $ds = 0$.) The presence of the first effect, which consists in a displacement of the spectral lines towards the red end, has not yet been definitely established. Whereas the efforts to detect this shift in the gravitational field of the sun have so far been fruitless, observations of the spectra of other fixed stars seem to indicate with great probability that it actually exists. The second effect, however, viz. the deflection of light by gravitation was established beyond doubt on May 29, 1919, on the occasion of the total eclipse of the sun. The light from a star which, on its way to the earth, passes close by the sun, is attracted by the latter's intense gravitational field. This should, according to theory, express itself in an apparent displacement of the star. Since these stars which happen to be near the sun (as projected on the celestial sphere) are only visible to the eye or a photographic plate during a total eclipse of the sun, this inference from theory can only be tested upon such occasions. Two expeditions were sent out from England to observe the above eclipse. They succeeded in finding that the displacement of the apparent position of these stars was actually such as had been prophesied by Einstein, and, indeed, to the exact amount he had previously calculated. This confirmation is doubtless one of the most brilliant achievements of human thought, and, in its theoretical significance, even surpasses the famous discovery of the planet Neptune from the calculation of Leverrier and Adams. The general theory of relativity has in this way successfully undergone the severest tests. The world of science pays homage to the triumphant power with which the correctness of the physical content of the theory and the truth of its philosophical foundations are confirmed by experience.

The assertion that *all* motions and accelerations are relative is equivalent to the assertion that space and time have no physical objectivity. One statement comprehends the other. Space and time are not measurable in themselves:

they only form a framework into which we arrange physical events. As a matter of principle, we can choose this framework at pleasure; but actually we do so in such a way that it conforms most closely to observed events (e.g. so that the 'geodetic lines' of the framework assume a distinctive physical rôle); we thus arrive at the simplest formulation of physical laws. An order has no independent existence, but manifests itself only in ordered things. Minkowski had as a result of the special theory of relativity enunciated the proposition in terse language (perhaps not wholly free from criticism) that space and time *in themselves* are reduced to the status of mere shadows, and only an indissoluble synthesis of both has an independent existence. So, on the basis of the general theory of relativity, we may now say that this synthesis itself has become a mere shadow, an abstraction; and that only the oneness of space, time, and things has an independent existence.

IX

THE FINITUDE OF THE UNIVERSE

In Newton's mechanics, and, indeed, in pre-Einsteinian physics altogether, space played a part which was altogether independent of any considerations about matter. Just as a vessel can exist free of content and preserve its form, space was to preserve its properties, whether 'occupied' by matter or not. The general theory of relativity has taught us that this view is groundless and misleading. 'Space', according to *it,* is possible only when matter is present, which then determines its physical properties.

This standpoint, which arises out of the general theory of relativity, is proved to be the only justifiable one, when we approach the cosmological question of the structure of the universe as a whole. Certain difficulties had already been encountered earlier, which clearly showed that Newton's cosmology was untenable; but it never suggested itself to anyone that Newton's doctrine of space might be partly responsible for these difficulties. The relativity theory yields an unexpected and wondrous solution of the discrepancies, which is of exceeding importance for our picture of the world.

It was generally believed by the ancients that the cosmos was bounded by a mighty sphere, to the inner surface of which the fixed stars were thought to be attached in some way. Even Copernicus did not succeed in destroying this belief. He had placed the sun in the middle of the planets moving around it, and recognized the earth as one planet amongst many others, but not yet the sun as one of many fixed stars. In comparison with this naïve view, the picture of the world must have seemed to become both enriched and exalted when Giordano Bruno propounded the doctrine of the infinity of the worlds in space. It was alluring to the imagination to think of the innumerable stars as being also suns similar to our own, and poised in space, and of space as extending to infinity, not limited by any rigid sphere, nor enclosed by any 'crystal dome'. Bruno glorifies the freedom of spirit which emanates from this extension of the world system in rapturous lines:

> Now unconfined the wings stretch out to heaven,
> Nor shrink beneath a crystal firmament
> Aloft into the aether's fragrant deeps,
> Leaving below the earth-world with its pain,
> And all the passions of mortality.

Up to the present day the conception of the world as a whole described in these lines has had complete sway. It was certainly, from an aesthetic standpoint, most attractive and most satisfactory for the philosopher to picture the cosmos as composed of the world of matter infinitely extended into infinite space; a traveller on the way to infinitely distant regions meets with ever new stars, even if he continue through all eternity, without reaching the limits of their realms or exhausting their number. It is true that the stars have been sown with great scarcity in the heavenly regions; a comparatively small amount only of matter is scattered over a great volume of space; but its *mean* density is to be the same everywhere, and is not to become zero even at infinity. So that, if we fix upon a certain amount of mass in some great volume of celestial space, and divide it by the size of this volume, we should by choosing a continually larger volume arrive at a constant finite value for the mean density. From the point of view of natural philosophy, such a picture of the world would be highly satisfactory. It would have neither beginning nor end, neither a centre nor boundaries, and space would nowhere be empty.

But the celestial mechanics of Newton is *incompatible* with this view. For, if we assume the strict validity of Newton's gravitational formula, according to which masses exert a mutually attractive force varying inversely as the square of the distance, calculation shows that the effects at a certain point of an infinite number of masses present at infinite distances, according to the above view, do not sum up to a certain finite gravitational force at the point, but that only infinite and indeterminate values are obtained.

Einstein proves this in an elementary way as follows: If ρ is the average density of matter in the universe, then the amount of matter contained in a large sphere of radius R is $4/3 \, \pi\rho \, R^3$. The same expression (by a familiar theorem of the theory of potential) gives the number of 'lines of force', due to gravitation which pass through the surface of the sphere. The extent of this surface is $4\pi R^2$, so that there are $\frac{1}{3}\rho R$ lines of force to every unit area of surface. But this latter number expresses the intensity of the force which is exerted by the gravitational effect of the contents of the sphere at a point on the surface: it clearly becomes infinite, if R increases beyond all limits.

As this is impossible, the universe cannot, on Newton's theory, be constituted as was just portrayed; gravitational potential must become zero at infinity, and the cosmos must present the picture of an island of finite extent surrounded on all sides by infinite 'empty space': and the mean density of matter would be infinitely small.

But such a picture of the universe would be unsatisfactory to the highest degree. The energy of the cosmos would constantly decrease, as radiation would disappear into infinite space; and matter, too, would gradually disperse. After a certain time the world would have died an inglorious death.

Now these exceedingly awkward consequences are inseparably connected with Newton's theory. The astronomer Seeliger, who laid bare these shortcomings to their full extent, sought to escape them by assuming that the attractive force between two masses decreased more rapidly than Newton's law demands. With the help of this hypothesis, he actually succeeds in maintaining without contradiction this idea of a world infinitely extended, filling all space with matter of a mean density. An unsatisfactory feature of this theory is, however, contained in the fact that the hypothesis is invented *ad hoc,* and is not occasioned or supported by any other experience.

Great interest thus circles round the question whether it is not possible to solve the cosmological problem by some new theory which is entirely satisfactory in every way. The suggestion forces itself upon us that the general theory of relativity might be able to do this; for, in the first place, it gives us information about the nature of gravitation towards which the Newtonian law represents only an approximation; secondly, it sheds an entirely new light on the problem of space. We have therefore reason for hoping that it will give us important disclosures about the question of the finitude of the world in space.

When Einstein investigated whether his theory was to be brought into closer harmony with the assumption of an infinite world with an average uniform density of distribution of stars than had been possible for Newton's theory, he first met with disappointment. For it appeared that a universe constructed in accordance with the hopes expressed above was just as little compatible with the new mechanics as with that of Newton.

As we know, the space of the new theory of gravitation is not Euclidean in structure, but departs somewhat from this shape, conforming in its measure-relations to the distribution of matter. Now if it were possible that, corresponding to the world-picture of Giordano Bruno, a uniform distribution of stars on the average existed for infinite space, then, in spite of deviations in particular places, space could still roughly be called Euclidean as a whole: just as I might call the ceiling of my room plane, by forming an abstraction which neglects the little roughnesses of its surface. Calculation, however, shows that such a structure of space—Einstein calls it quasi-Euclidean—is *not* possible in the general theory of relativity. On the contrary, according to this theory, the mean density of matter must necessarily be zero in infinite non-Euclidean space; i.e. we are again driven to the world-system which was discussed above, which would consist of a finite aggregation of matter in otherwise empty space of infinite dimensions.

This view, which was unsatisfactory for Newton's theory, is still more so for the general theory of relativity. Not only do the objections which were pointed out above apply in this case also, but new ones arise in addition. For, if we seek to find the mathematical boundary conditions for the quantities g at infinity, which correspond to this case, Einstein shows that we may attempt it in two ways. We might, in the first place, think of assigning to the g's the same boundary values which are allotted to them at infinity in the mathematical treatment of planetary motions. For the planetary system certain limiting values ($g_{11} = g_{22} = g_{33} = -1$, $g_{44} = +1$, the other g's $= 0$) are permissible, since we have still to take into account the presence of the stellar system at great distances; but the extension of this method to the whole universe is incompatible with the fundamental ideas of the relativity theory in two respects. First, a perfectly definite choice of co-ordinate systems would be imperative for this; and second, the inertial mass of a body would, contrary to our hypotheses, no longer be solely due to the presence of other bodies; but a material point would still possess inertial mass if it were at an infinite distance from other bodies, or even if it were entirely isolated and left alone in the world-space. This is contrary to the trend of thought of the general principle of relativity; and we see that only those solutions come into consideration in which the inertia of a body vanishes at infinity.

Einstein now showed (and this appeared to be the second way) that one might indeed assume boundary conditions for the g's at infinity, which would fulfil the latter demand; and that a world-picture drawn in this way would even have an advantage over the Newtonian one, inasmuch as no star and no ray of light, according to it, could disappear in infinite space, but would finally have to return into the system. But he also showed that such boundary conditions would be in absolute disagreement with the actual state of the stellar system, as experience presents it to us. The gravitational potentials would have to increase at infinity beyond all limits, and very great relative velocities of the stars would necessarily occur, whereas, in fact, we observe that the motions of the stars take place extremely slowly compared with the velocity of light. The fact of the small velocity of the stars is indeed one of the most striking peculiarities, common to all members of the stellar system, which offer themselves to observation, and can be used as a basis for cosmological speculations. In virtue of this property, we can unhesitatingly regard the matter in the cosmos as at rest to a first approximation (if we choose an appropriate system of reference); we consequently base our calculations on this assumption.

We thus find that the second method likewise does not

lead to the goal. The inference is that, according to the relativity theory, the universe cannot be a finite complex of stars existing in infinite space; this, after the above remarks, means that we cannot regard space as quasi-Euclidean. What possibility now remains?

At first it seemed as if no reply was forthcoming from the theory; but Einstein soon discovered that it was still possible to generalize his original gravitational equations slightly further. After this small extension of the formulae, the general theory of relativity has the inestimable advantage of giving us an unmistakable answer, whereas the previous Newtonian theory left us in total uncertainty, and could only rescue us from forming a highly undesirable picture of the universe by making new and unconfirmed hypotheses.

If we again suppose the matter of the universe to be distributed with absolutely uniform density and to be at rest, the calculation leaves no doubt but that space is *spherical* in structure (there is the additional possibility that it might be 'elliptical' in constitution, but we may neglect this case, which seems to be of mathematical rather than of physical interest). Since matter does not actually occupy space uniformly and is not at rest, but only shows the same density of distribution *as a mean,* we must regard space as quasi-spherical (i.e. on the whole it is spherical, but departs from this form in its smaller parts, just as the earth is only an ellipsoid as a whole, but is, when considered in smaller portions, possessed of an irregularly formed surface).

What the term 'spherical space' is intended to convey is probably known to the reader through Helmholtz's popular essays. He, as we know, describes the three-dimensional analogy to a spherical surface; the former has, like the latter, the property of being circumscribed, i.e. it is unlimited and yet finite. The comparison with the surface of a sphere must not mislead one to confuse in one's mind 'spherical' with sphere-shaped. A sphere is bounded by its surface, the latter cutting it out of space as a part of it; spherical space, however, is not a part of infinite space, but has simply no limits. If I start out from a point of our spherical world and continually proceed along a 'straight line', I shall never reach a limiting surface; the 'crystal dome', which according to the ancients was supposed to encompass the universe, exists just as little for Einstein as it did for Giordano Bruno. There is no space outside the world; space exists only in so far as matter exists, for space in itself is merely a product of abstraction. If, from any point, we draw the straightest lines in all directions, these at first, of course, diverge from one another, but then approach again, in order finally to meet at one point as before. The totality of such lines fills the world-space entirely, and the volume of the latter is finite. Einstein's theory even enables us to calculate its numerical value for a given density of distribution; we thus obtain the volume

$$V = \frac{7.10^{41}}{\sqrt{p^3}}$$ cubic centimetres—an enormously high figure;

for p, the mean density of matter, has an exceedingly small value. The structure of the universe, which the general theory of relativity unveils to us, is astounding in its logical consistency, imposing in its grandeur, and equally satisfying for the physicist as for the philosopher. All the difficulties which arose from Newton's theory are overcome; yet all the advantages which the modern picture of the world presents, and which elevate it above the view of the ancients, shine with a clearer lustre than before. The world is not confined by any boundaries, and is yet harmoniously complete in itself. It is saved from the danger of becoming desolate, for no energy or matter can wander off to infinity, because space is *not* infinite. The infinite space of the cosmos has certainly had to be rejected; but this does not signify such sacrifice as to reduce the sublimity of the picture of the world. For that which causes the idea of the infinite to inspire sublime feelings is beyond doubt the idea of the endlessness of space (actual infinity could not in any case be imagined); and this absence of any barrier, which excited Giordano Bruno to such ecstasy, is not infringed in any way.

By a combination of physical, mathematical, and philosophic thought genius has made it possible to answer, by means of exact methods, questions concerning the universe which seemed doomed for ever to remain the objects of vague speculation. Once again we recognize the power of the theory of relativity in emancipating human thought, which it endows with a freedom and a sense of power such as has been scarcely attained through any other feat of science.

X

RELATIONS TO PHILOSOPHY

It is scarcely necessary to mention that the words space and time in the preceding chapters have been used only in the 'objective' sense in which these conceptions occur in natural science. 'Subjective' psychological experience of extension in space and order in time is quite distinct from these.

Ordinarily there is nothing to induce us to analyse this difference in detail; the physicist does not need to concern himself in the slightest with the investigations of the psychologist into spatial perception. But when we wish to form a clear picture of the ultimate epistemological foundations of natural science, it becomes necessary to give an adequate account of the relationship between these two points of view. This is the task of the philosopher; for it is generally accepted that it is for philosophy to reveal the fundamental assumptions of the separate sciences, and bring them into harmony with one another.

What leads us to speak of space and time at all? What is the psychological source of these notions? There is no doubt that all our perceptions of space, and the conclusions resulting therefrom, emanate from certain properties of our sense-impressions, viz. from those properties which we term 'spatial' and which do not allow of closer definition: for we get our knowledge of them only from direct experience. Just as it is impossible for me to explain to a person who has been born blind, by means of a definition in words, what I experience when I see a green surface, so it is impossible for me to describe what is meant when I ascribe to this green appearance a definite extension and position in the field of vision. In order to know what is meant, we must be able to 'behold' it: we must have visual perceptions or impressions. This spatial quality, which is an essential accompaniment of visual impressions, is thus *intuitive* ('*anschaulich*'). We assign the term in an extended sense to all the other data of our world of presentations and perceptions, not only the visual ones. The perceptions of the other senses, more particularly the tactual and kinaesthetic (muscular and articular) presentations, have properties which we likewise term '*spatial*'. In fact the intuition which the blind have of space consists, exclusively, of such data. A sphere feels different from a cube to the touch: I experience different muscular sensations in the arm, according as I describe with my hand a long or a short, a gently curved or a zigzag line. These differences constitute the space quality (*Räumlichkeit*) of the tactual and muscular perceptions: it is these that the person born blind has in his mind when he hears of different localities or dimensions.

The data, however, of the various realms of perception cannot be compared with one another (e.g. the space arising out of tactual presentations is entirely dissimilar in kind from that of the optical presentations: a man born blind, who has a knowledge of the first only, cannot, from it, form any notion of the latter). Tactual space has so far not the slightest resemblance to visual space, and the psychologist finds himself obliged to say that there are just as many spaces for our intuition as we have senses.

The space of the physicist, however, which we set up as objective in opposition to these subjective spaces, is a single definite one, and we think of it as independent of our sense impressions (but of course not independent of physical objects; on the contrary, it is only real in conjunction with

them). It is not identical with any of the above spaces of
intuition, for it has quite different properties. If we look at
a rigid cube, for instance, we find that its form changes for
our visual sense according to the side at which, or the distance from which, we view it. The apparent length of its
edges varies, and yet we ascribe to it the same physical
shape. We get a similar result, in forming a judgment about
a cube, by means of our sense of touch: by which we also receive entirely different impressions, according as we touch
larger or smaller parts of its surface, or according to
the parts of the skin which come in contact with it; yet in
spite of these different impressions we pronounce the cubical
form of the object to have remained unaltered. The objects
of physics are therefore *not* the data of sense: the space of
physics is not in any way given with our perceptions, but is
a product of our conceptions. We cannot therefore ascribe
to physical objects the space of intuition with which our visual perceptions have made us acquainted, nor that which we
find present in our tactual presentations, but only a conceptual arrangement, which we then term objective space, and
determine by means of a suitably disposed manifold of numbers (co-ordinates). Hence we see that the same thing holds
true for intuitional space as for other qualities of the sense-
data such as tones, colours, &c. Physics does not know col-
our as a property of the object with which it is associated,
but only frequencies of the vibrations of electrons. It has
no knowledge of qualities of heat, but only of kinetic energy
of the molecules.

Similar arguments apply in the consideration of subjective psychological time. A special psychological time cannot indeed be claimed for the realm governed by each particular sense; for it is one and the same time-character which
permeates all experiences—not only those of the senses—in
the same way. This direct experience of duration, of earlier

and later, is nevertheless an ever-changing intuitional factor, which makes one and the same objective event appear, according to mood and attentiveness, now long, now short: a factor which vanishes altogether during sleep, and bears an entirely different stamp according to the wealth of associations of the experience. In short, it is easily distinguishable from physical time, which only signifies an arrangement having the properties of a one-dimensional continuum. This objective order or arrangement has just as little to do with the intuitional experience of duration as the three-dimensional order of objective space has to do with the intuitional experiences of extension, as presented optically or tactually. In recognizing this, we get the pith of Kant's doctrine of the 'Subjectivity of Time and Space', according to which both are merely 'forms' of our intuition, and cannot be ascribed to the 'things-in-themselves'. Kant himself does not give clear expression to this truth; for he always talks of 'space' only, without drawing a dividing line between the intuitional spaces of the various senses, or between them and the space of bodies as implied in physics. Instead of this, he merely opposes the unknowable arrangement of the 'things-in-themselves' to the space and time of the things as given by the senses. We, on the other hand, find occasion to distinguish only between the intuitional psychological spaces and non-intuitional physical space. Just because the latter is purely conceptual, it is quite impossible —contrary to the opinion of many a follower of Kant—for intuition to give us the slightest information as to whether physical space is Euclidean or not. In conjunction with objective time, physical space is designated by the four-dimensional scheme which we have repeatedly discussed above, and which in mathematical language can simply be treated as the manifold of all number quadruples x_1, x_2, x_3, x_4. In this objective scheme there is no distinction between a 'time' distance and a 'space' distance. This is the point which receives full recognition for the first time through the theory of relativity. Both simply appear as one-dimensional continua; and there is no room left in this conception for the intuitional difference between duration (length of time) and extension (length of space). It does not matter how fundamental a part this difference plays for consciousness.

It is obvious that in the first instance only the intuitional psychological spaces and times are given us; and we must inquire how we have, by starting from them, arrived at the construction of the objective space-time manifold. This construction is not indeed a product of natural science, but is a necessity of our daily life; for when we ordinarily talk of the position and shape of bodies, we are always already thinking of physical space, which is conceived as independent of individuals and of the organs of sense. Of course, we always represent to our consciousness shapes and distances, about which we are thinking, by visual and tactual means and kinaesthetic presentations: because we always strive, as far as possible, to exhibit non-intuitional conceptual relations in our thinking by sensory substitutes which may act as their representatives, but are no more than sense-representatives of the physical conception of space. The former are not to be confused with the latter, nor must they lead one to regard the latter with Kant as likewise intuitional.

The answer to our question, as to the genesis of the physical conception of space from the intuitional data of the psychological spaces, is now quite plain. These spaces are essentially dissimilar and incapable of comparison with one another; but they have, as our experiences teach us, a perfectly definite uniform functional relation to one another. Tactual perceptions, e.g., correlate themselves with visual perceptions. A certain correspondence exists between the two spheres; and through this correspondence it is possible to arrange all spatial perceptions into one scheme, this being just what we call objective space. If in feeling over an object my skin nerves receive a perception-complex of the 'cube form', I can, by adopting proper measures (lighting a candle, opening my eyes, &c.), receive certain visual perception complexes, which I likewise designate as 'cube form'.[1] The optical impression is *toto caelo* different from the tactual one; but experience teaches me that they go hand in hand with one another. In the case of persons born blind, who acquire the sight of their eyes through an operation, we have an opportunity of studying their gradual training in associating the data of the two realms of sight and touch.[2]

Now it is important to understand quite clearly what particular experiences lead us to connect a perfectly definite element of optical space with a perfectly definite element of tactual space, and thereby to form the conception of a ' point ' in objective space. For it is here that experiences arising out of coincidences come into account. In order to fix a point in space, we must in some way or other, directly or indirectly, *point* to it: we must make the point of a pair of compasses, or a finger, or the intersection of cross-wires, coincide with it (i.e. bring about a time-space coincidence of two elements which are usually apart). Now these coincidences always occur consistently for all the intuitional spaces of the various senses and for various indi-

---

[1] Vide Locke's *Essay on Human Understanding*, bk. ii, ch. 9, s. 8.

[2] This view is familiar to the English reader from Berkeley's *New Theory of Vision*. (Fraser, Oxford edition, vol. i.) Cf. Dufaur, *Archives des sciences physiques et naturelles*, tome 58, p. 232.

Schopenhauer cited various instances in chap. iv of his *Fourfold Root of the Principle of Sufficient Reason*, mentioning in partcular Cheselden's blind man, a case recorded in *Phil. Trans.* vol. 35 (Trans.).

viduals. It is just on account of this that a ' point ' is
defined which is objective, i.e. independent of individual experiences and valid for all.   An extended pair of compasses applied to the skin excites two sensations of pricking; but if I bring the two points together so that they
occupy the same spot in optical space, I only get *one* sen-
sation of pricking, and there is also coincidence in tactual space.   Upon close investigation, we find that we arrive
at the construction of physical space and time by just this
method of coincidences and by no other process.   The
space-time manifold is neither more nor less than the quintessence of objective elements as defined by this method.
The fact of its being a four-dimensional manifold follows
from experience in the application of the method itself.

This is the outcome of our analysis of the conceptions of
space and time; it is an analysis of psychological data
regarded as our sources of knowledge.   We see that we
encounter just that significance of space and time which
Einstein has recognized to be essential and unique for
physics, where he has established it in its full right.
For he rejected Newton's conceptions, which denied
the origin we have assigned to them, and founded physics
on the conception of the coincidence of events.   Here we
have the realization of an eminently desirable point of contact between physical theory and the theory of knowledge.

In one matter physical theory goes far beyond the bounds
within which we have psychological data.   Physics intro-
duces, as its ultimate indefinable conception, the coincidence
of two *events;* on the other hand, the psycho-genetic analysis of the idea of objective space ends in the conception of
the space-time coincidence of two *elements of perception.*
Are they to be regarded simply as one and the same thing?

Rigorous positivism, such as that of Mach, affirms them
to be so.   According to him, the directly experienced ele-

ments such as colours, tones, pressures, warmths, &c., are the sole reality, and there are no other actual events beyond the coming and going of these elements. Wherever else in physics other coincidences are mentioned, they are only abbreviated modes of speech, economical working-hypotheses, not realities as perceptions are. Looked at from this point of view, the conception of the physical world in its objective four-dimensional scheme would merely be an abridged statement of the correspondence of the subjective time-space experiences in the realms of the various senses, and *nothing more.*

This view is, however, not the only possible interpretation of scientific facts. If distinguished investigators in the domain of the exact sciences do not cease to urge that the picture of the world as offered by Mach fails to satisfy them, the ground for it is doubtless to be sought in this, that the quantities which occur in physical laws do not all indicate 'elements' in Mach's sense.[1] The coincidences which are expressed by the differential equations of physics are not immediately accessible to experience. They do not directly signify a coincidence of sense-data; they denote non-sensory magnitudes, such as electric and magnetic intensities of field and similar quantities. There is no argument whatsoever to force us to state that only the intuitional elements, colours, tones, &c., exist in the world. We might just as well assume that elements or qualities which cannot be directly experienced also exist. These can likewise be termed 'real', whether they be comparable with intuitional ones or not. For example, electric forces can just as well signify elements of reality as colours and tones. They are *measurable,* and there is no reason why epistemology should reject the criterion for reality which is used

[1] The English reader will find the corresponding theory in K. Pearson, *Grammar of Science.*

in physics (v. p. 21). The conception of an electron or an atom would then not necessarily be a mere working hypothesis, a condensed fiction, but could equally well designate a real connexion or complex of such objective elements: just as the conception of the 'ego' denotes a real complex of intuitional elements. The picture of the world, as presented by physics, would then be a system of symbols arranged into a four-dimensional scheme, by means of which we get our knowledge of reality; that is, *more* than a mere auxiliary conception, allowing us to find our way through given intuitional elements.

The two views stand in opposition; and I believe that there is no rigorous proof of the correctness of the one and the falseness of the other. The reasons which induce me to declare myself in favour of the second—which may, in contrast to the strictly positivist view, be called realistic —are as follows:—

First, it seems to me to be purely arbitrary, nay, dogmatic, to allow only the intuitional elements and their relationships to be valid as *realities*. Why should intuitional experiences be the only 'events' in our world? Why should there not be other events besides these?

We find that the processes of science do not justify us in thus narrowing the conception of reality. It was put forward in opposition to certain fallacious metaphysical views; but these can be avoided in other ways.

Secondly, the strictly positivist picture of the world seems to me to be unsatisfactory on account of a certain lack of continuity. In narrowing down the conception of reality in the above sense, we tear, as it were, certain holes in the fabric of reality, which are patched up by mere auxiliary conceptions. The pencil in my hand is to be regarded as real, whereas the molecules which compose it are to be pure fictions. This antithesis, often uncertain and fluctuating, between conceptions which denote something real and those which are only working-hypotheses, finally becomes unbearable. It is avoided by the assumption, which is certainly allowable, that every conception which is actually of use for a description of physical nature can likewise be regarded as a sign of something real. I believe that, in striving to illuminate even the innermost recesses of the theory of knowledge, we need never give up this assumption, and that it renders possible a view of the world harmonious in its last details and perfect in itself, which also satisfies the demands imposed upon thought by the realist's attitude of mind, but without making it necessary to give up any of the advantages of the positivist view of the world.

One of its chief advantages is that the relation of the separate theories to one another receives due recognition and a proper measure of value. We felt ourselves impelled several times in the course of the discussion to explain clearly to ourselves that, in many cases, there is no possibility, and no urgent need, to distinguish one point of view from the others as the only *true* one. It can never be proved that Copernicus alone is in the right, and that Ptolemy is wrong. There is no logical ground which can compel us to set up the theory of relativity as the only true one in opposition to the absolute theory, or to declare that the Euclidean determinations of measure are merely right or wrong. The most that can be done is to show that, of these alternatives, the one view is simpler than the other, and leads to a more finished, more satisfactory picture of the world.

Every theory is composed of a network of conceptions and judgments, and is *correct* or *true* if the system of judgments indicates the world of facts *uniquely*. For, if such a unique correspondence exists between conceptions and

reality, it is possible, with the assistance of the network of
judgments in the theory, to derive the successive steps in
the phenomena of nature, e.g. to predict occurrences in the
future. And the fulfilment of such prophecies, the agreement between calculation and observation, is the only
means of proving that a theory is true. It is, however, possible to indicate identically the *same* set of facts by means
of *various* systems of judgments; and consequently there
can be various theories in which the criterion of truth is
equally well satisfied, and which then do equal justice to
the observed facts, and lead to the same predictions. They
are merely different systems of symbols, which are allo-
cated to the same objective reality: different modes of expression which reproduce the same set of facts. Amongst
all the possible views which contain the same nucleus of
truth in this way, there must be one which is simplest; and
our reason for preferring just this one is not founded upon
reasons of practical economy, a sort of mental indolence (as
has been held by some). There is a logical reason for it, inasmuch as the simplest theory contains a minimum number
of *arbitrary* factors. The more complicated views necessarily contain superfluous conceptions, of which I can dis-
pose at pleasure, and which are consequently not condi-
tioned by the facts under consideration; about which, therefore, I am right in asserting that nothing real corresponds
to them, regarded apart from the other conceptions. In the
case of the simplest theory, on the other hand, the rôle of
each particular conception is made imperative by the facts:
such a theory forms a system of symbols, all of them in-
dispensable. Lorentz's aether-theory (v. p. 10), for
example, declares one co-ordinate system to be unique
among all others, but does not essentially afford the means
of ever actually specifying this system. His theory is thus
encumbered with the conception of absolute motion,

whereas the conception of relative motion suffices for a unique description of the facts. The former is never capable of application alone, but only in certain combinations, which are embraced in the conception of relative motion.

Now, the conceptions of space and time, in the form in which they have hitherto occurred, in physics are included among these superfluous factors. This we have recognized as a result of the general theory of relativity. They, too, cannot be applied separately; but only in so far as they enter into the conception of the space-time coincidence of events. We may therefore reiterate that only in this union do they indicate something real, but not when taken alone.

We see how stupendous is the theoretical range of these new views. Einstein's analysis of the conceptions of space and time belongs to the same category of philosophic evolution as David Hume's criticism of the ideas of substance and causality. In what way this development will continue, we cannot yet say. The method which characterizes it is the only fruitful one for the theory of knowledge, consisting as it does in a searching criticism of the fundamental ideas of science, stripping off everything that is superfluous and with ever-increasing clearness exposing the ultimate pure content.

INDEX

A CATALOG OF SELECTED
DOVER BOOKS
IN SCIENCE AND MATHEMATICS

Physics

OPTICAL RESONANCE AND TWO-LEVEL ATOMS, L. Allen and J. H. Eberly. Clear, comprehensive introduction to basic principles behind all quantum optical resonance phenomena. 53 illustrations. Preface. Index. 256pp. 5⅜ x 8½.   65533-4

QUANTUM THEORY, David Bohm. This advanced undergraduate-level text presents the quantum theory in terms of qualitative and imaginative concepts, followed by specific applications worked out in mathematical detail. Preface. Index. 655pp. 5⅜ x 8½.   65969-0

ATOMIC PHYSICS: 8th edition, Max Born. Nobel laureate's lucid treatment of kinetic theory of gases, elementary particles, nuclear atom, wave-corpuscles, atomic structure and spectral lines, much more. Over 40 appendices, bibliography. 495pp. 5⅜ x 8½.   65984-4

A SOPHISTICATE'S PRIMER OF RELATIVITY, P. W. Bridgman. Geared toward readers already acquainted with special relativity, this book transcends the view of theory as a working tool to answer natural questions: What is a frame of reference? What is a "law of nature"? What is the role of the "observer"? Extensive treatment, written in terms accessible to those without a scientific background. 1983 ed. xlviii+172pp. 5⅜ x 8½.   42549-5

AN INTRODUCTION TO HAMILTONIAN OPTICS, H. A. Buchdahl. Detailed account of the Hamiltonian treatment of aberration theory in geometrical optics. Many classes of optical systems defined in terms of the symmetries they possess. Problems with detailed solutions. 1970 edition. xv+360pp. 5⅜ x 8½.   67597-1

PRIMER OF QUANTUM MECHANICS, Marvin Chester. Introductory text examines the classical quantum bead on a track: its state and representations; operator eigenvalues; harmonic oscillator and bound bead in a symmetric force field; and bead in a spherical shell. Other topics include spin, matrices, and the structure of quantum mechanics; the simplest atom; indistinguishable particles; and stationary-state perturbation theory. 1992 ed. xiv+314pp. 6⅛ x 9¼.   42878-8

LECTURES ON QUANTUM MECHANICS, Paul A. M. Dirac. Four concise, brilliant lectures on mathematical methods in quantum mechanics from Nobel Prize–winning quantum pioneer build on idea of visualizing quantum theory through the use of classical mechanics. 96pp. 5⅜ x 8½.   41713-1

THIRTY YEARS THAT SHOOK PHYSICS: The Story of Quantum Theory, George Gamow. Lucid, accessible introduction to influential theory of energy and matter. Careful explanations of Dirac's anti-particles, Bohr's model of the atom, much more. 12 plates. Numerous drawings. 240pp. 5⅜ x 8½.   24895-X

ELECTRONIC STRUCTURE AND THE PROPERTIES OF SOLIDS: The Physics of the Chemical Bond, Walter A. Harrison. Innovative text offers basic understanding of the electronic structure of covalent and ionic solids, simple metals, transition metals and their compounds. Problems. 1980 edition. 582pp. 6⅛ x 9¼.   66021-4

CATALOG OF DOVER BOOKS

QUANTUM MECHANICS: Principles and Formalism, Roy McWeeny. Graduate student–oriented volume develops subject as fundamental discipline, opening with review of origins of Schrödinger's equations and vector spaces. Focusing on main principles of quantum mechanics and their immediate consequences, it concludes with final generalizations covering alternative "languages" or representations. 1972 ed. 15 figures. xi+155pp. 5⅜ x 8½. 42829-X

INTRODUCTION TO QUANTUM MECHANICS WITH APPLICATIONS TO CHEMISTRY, Linus Pauling & E. Bright Wilson, Jr. Classic undergraduate text by Nobel Prize winner applies quantum mechanics to chemical and physical problems. Numerous tables and figures enhance the text. Chapter bibliographies. Appendices. Index. 468pp. 5⅜ x 8½. 64871-0

METHODS OF THERMODYNAMICS, Howard Reiss. Outstanding text focuses on physical technique of thermodynamics, typical problem areas of understanding, and significance and use of thermodynamic potential. 1965 edition. 238pp. 5⅜ x 8½. 69445-3

TENSOR ANALYSIS FOR PHYSICISTS, J. A. Schouten. Concise exposition of the mathematical basis of tensor analysis, integrated with well-chosen physical examples of the theory. Exercises. Index. Bibliography. 289pp. 5⅜ x 8½. 65582-2

THE ELECTROMAGNETIC FIELD, Albert Shadowitz. Comprehensive undergraduate text covers basics of electric and magnetic fields, builds up to electromagnetic theory. Also related topics, including relativity. Over 900 problems. 768pp. 5⅜ x 8¼. 65660-8

GREAT EXPERIMENTS IN PHYSICS: Firsthand Accounts from Galileo to Einstein, Morris H. Shamos (ed.). 25 crucial discoveries: Newton's laws of motion, Chadwick's study of the neutron, Hertz on electromagnetic waves, more. Original accounts clearly annotated. 370pp. 5⅜ x 8½. 25346-5

RELATIVITY, THERMODYNAMICS AND COSMOLOGY, Richard C. Tolman. Landmark study extends thermodynamics to special, general relativity; also applications of relativistic mechanics, thermodynamics to cosmological models. 501pp. 5⅜ x 8½. 65383-8

STATISTICAL PHYSICS, Gregory H. Wannier. Classic text combines thermodynamics, statistical mechanics, and kinetic theory in one unified presentation of thermal physics. Problems with solutions. Bibliography. 532pp. 5⅜ x 8½. 65401-X

Paperbound unless otherwise indicated. Available at your book dealer, online at **www.doverpublications.com**, or by writing to Dept. GI, Dover Publications, Inc., 31 East 2nd Street, Mineola, NY 11501. For current price information or for free catalogs (please indicate field of interest), write to Dover Publications or log on to **www.doverpublications.com** and see every Dover book in print. Dover publishes more than 500 books each year on science, elementary and advanced mathematics, biology, music, art, literary history, social sciences, and other areas.